How It Works

by

Archibald Williams

Double 9 BOOKS

How It Works

by Archibald Williams

ISBN: 978-93-58714-07-4

Published by

DOUBLE 9 BOOKS
2/13-B, Ansari Road
Daryaganj, New Delhi – 110002
info@double9books.com
www.double9books.com
Tel. 011-40042856

ABOUT THE AUTHOR

Archibald Williams (1886-1963) was a British author and historian. He is known for his works on military history, including "The Romance of War Inventions" (1916), "The Romance of Modern Invention" (1919), "The Romance of Aircraft" (1921), and "The Romance of the British Navy" (1925). Williams served in the British Army during World War I and his experience during the war heavily influenced his writing. He was also a prolific author of adventure stories and science fiction, writing under various pseudonyms such as "Archibald Williams", "William Wilson", and "Anthony Wingrave".

CONTENTS

PREFACE.

How does it work? This question has been put to me so often by persons young and old that I have at last decided to answer it in such a manner that a much larger public than that with which I have personal acquaintance may be able to satisfy themselves as to the principles underlying many of the mechanisms met with in everyday life.

In order to include steam, electricity, optics, hydraulics, thermics, light, and a variety of detached mechanisms which cannot be classified under any one of these heads, within the compass of about 450 pages, I have to be content with a comparatively brief treatment of each subject. This brevity has in turn compelled me to deal with principles rather than with detailed descriptions of individual devices—though in several cases recognized types are examined. The reader will look in vain for accounts of the Yerkes telescope, of the latest thing in motor cars, and of the largest locomotive. But he will be put in the way of understanding the essential nature of *all* telescopes, motors, and steam-engines so far as they are at present developed, which I think may be of greater ultimate profit to the uninitiated.

While careful to avoid puzzling the reader by the use of mysterious phraseology I consider that the parts of a machine should be given their technical names wherever possible. To prevent misconception, many of the diagrams accompanying the letterpress have words as well as letters written on them. This course also obviates the wearisome reference from text to diagram necessitated by the use of solitary letters or figures.

I may add, with regard to the diagrams of this book, that they are purposely somewhat unconventional, not being drawn to scale nor conforming to the canons of professional draughtsmanship. Where advisable, a part of a machine has been exaggerated to show its details. As a rule solid black has been preferred to fine shading in sectional drawings, and all unnecessary lines are omitted. I would here acknowledge my indebtedness to my draughtsman, Mr. Frank Hodgson, for his care and industry in preparing the two hundred or more diagrams for which he was responsible.

Four organs of the body—the eye, the ear, the larynx, and the heart—are noticed in appropriate places. The eye is compared with the camera, the larynx with a reed pipe, the heart with a pump, while the ear fitly opens the chapter on acoustics. The reader who is unacquainted with physiology will thus be enabled to appreciate the better these marvellous devices, far more marvellous, by reason of their absolutely automatic action, than any creation of human hands.

A.W.

Uplands, Stoke Poges, Bucks.

CHAPTER I.
THE STEAM-ENGINE.

What is steam?—The mechanical energy of steam—The boiler—The circulation of water in a boiler—The enclosed furnace—The multitubular boiler—Fire-tube boilers—Other types of boilers—Aids to combustion—Boiler fittings—The safety-valve—The water-gauge—The steam-gauge—The water supply to a boiler.

WHAT IS STEAM?

IF ice be heated above 32° Fahrenheit, its molecules lose their cohesion, and move freely round one another—the ice is turned into water. Heat water above 212° Fahrenheit, and the molecules exhibit a violent mutual repulsion, and, like dormant bees revived by spring sunshine, separate and dart to and fro. If confined in an air-tight vessel, the molecules have their flights curtailed, and beat more and more violently against their prison walls, so that every square inch of the vessel is subjected to a rising pressure. We may compare the action of the steam molecules to that of bullets fired from a machine-gun at a plate mounted on a spring. The faster the bullets came, the greater would be the continuous compression of the spring.

THE MECHANICAL ENERGY OF STEAM.

If steam is let into one end of a cylinder behind an air-tight but freely-moving piston, it will bombard the walls of the cylinder and the piston; and if the united push of the molecules on the one side of the latter is greater than the resistance on the other side opposing its motion, the piston must move. Having thus partly got their liberty, the molecules become less active, and do not rush about so vigorously. The pressure on the piston decreases as it moves. But if the piston were driven back to its original position against the force of the steam, the molecular activity—that is, pressure—would be restored. We are here assuming that no heat has passed through the cylinder or piston and been radiated into the air; for any loss of heat means loss of energy, since heat *is* energy.

THE BOILER.

The combustion of fuel in a furnace causes the walls of the furnace to become *hot,* which means that the molecules of the substance forming the walls are thrown into violent agitation. If the walls are what are called "good conductors" of heat, they will transmit the agitation through them to any surrounding substance. In the case of the ordinary house stove this is the air, which itself is agitated, or grows warm. A steam-boiler has the furnace walls surrounded by water, and its function is to transmit molecular movement (heat, or energy) through the furnace plates to the water until the point is reached when steam generates. At atmospheric pressure—that is, if not confined in any way—steam would fill 1,610 times the space which its molecules occupied in their watery formation. If we seal up the boiler so that no escape is possible for the steam molecules, their motion becomes more and more rapid, and *pressure* is developed by their beating on the walls of the boiler. There is theoretically no limit to which the pressure may be raised, provided that sufficient fuel-combustion energy is transmitted to the vaporizing water.

To raise steam in large quantities we must employ a fuel which develops great heat in proportion to its weight, is readily procured, and cheap. Coal fulfils all these conditions. Of the 800 million tons mined annually throughout the world, 400 million tons are burnt in the furnaces of steam-boilers.

A good boiler must be—(1) Strong enough to withstand much higher pressures than that at which it is worked; (2) so designed as to burn its fuel to the greatest advantage.

Even in the best-designed boilers a large part of the combustion heat passes through the chimney, while a further proportion is radiated from the boiler. Professor John Perry[1] considers that this waste amounts, under the best conditions at present obtainable, to eleven-twelfths of the whole. We have to burn a shillingsworth of coal to capture the energy stored in a pennyworth. Yet the steam-engine of to-day is three or four times as efficient as the engine of fifty years ago. This is due to radical improvements in the design of boilers and of the machinery which converts the heat energy of steam into mechanical motion.

CIRCULATION OF WATER IN A BOILER.

If you place a pot filled with water on an open fire, and watch it when it boils, you will notice that the water heaves up at the sides and plunges down at the centre. This is due to the water being heated most at the sides, and therefore being lightest there. The rising steam-bubbles also carry it up. On reaching the surface, the bubbles burst, the steam escapes, and the water loses some of its heat, and rushes down again to take the place of steam-laden water rising.

Fig. 1.

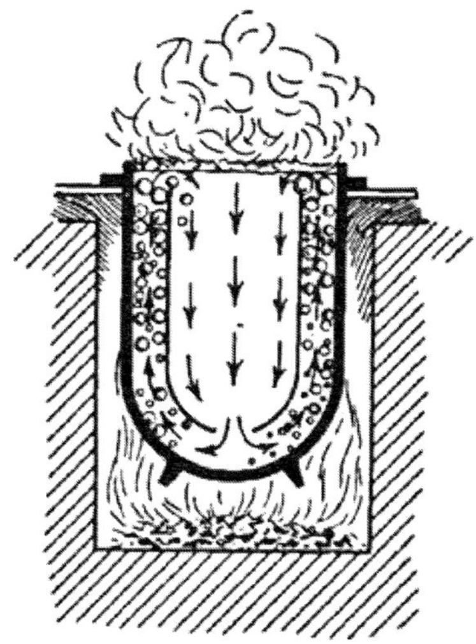

Fig. 2.

If the fire is very fierce, steam-bubbles may rise from all points at the bottom, and impede downward currents (Fig. 1). The pot then "boils over."

Fig. 2 shows a method of preventing this trouble. We lower into our pot a vessel of somewhat smaller diameter, with a hole in the bottom, arranged in such a manner as to leave a space between it and the pot all round. The upward currents are then separated entirely from the downward, and the fire can be forced to a very much greater extent than before without the water boiling over. This very simple arrangement is the basis of many devices for producing free circulation of the water in steam-boilers.

We can easily follow out the process of development. In Fig. 3 we see a simpleU-tube depending from a vessel of water. Heat is applied to the left leg, and a steady circulation at once commences. In order to increase the heating surface we can extend the heated leg into a long incline (Fig. 4), beneath which three lamps instead of only one are placed. The direction of the circulation is the same, but its rate is increased.

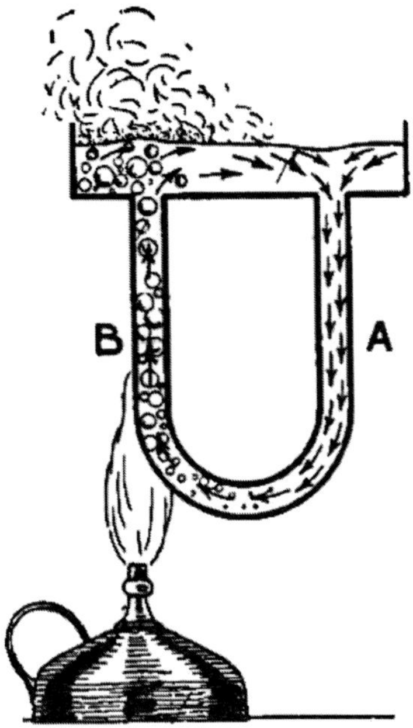

Fig. 3.

A further improvement results from increasing the number of tubes (Fig. 5), keeping them all on the slant, so that the heated water and steam may rise freely.

THE ENCLOSED FURNACE.

Fig. 4.

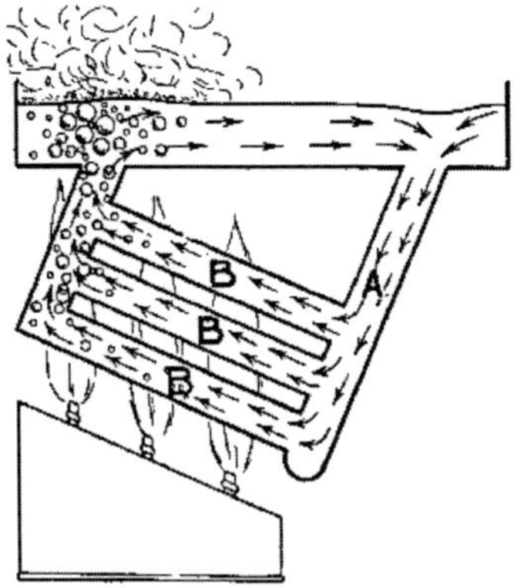

Fig. 5.

Still, a lot of the heat gets away. In a steam-boiler the burning fuel is enclosed either by fire-brick or a "water-jacket," forming part of the boiler. A water-jacket signifies a double coating of metal plates with a space between, which is filled with water (see Fig. 6). The fire is now enclosed much as it is in a kitchen range. But our boiler must not be so wasteful of the heat as is that useful household fixture. On their way to the funnel the flames and hot gases should act on a very large metal or other surface in contact with the water of the boiler, in order to give up a due proportion of their heat.

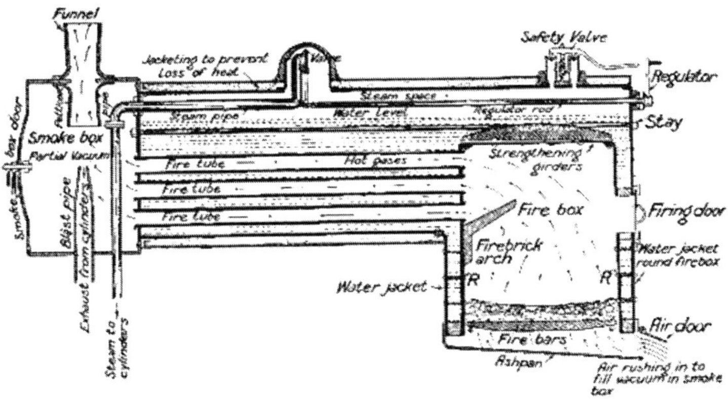

Fig. 6. —Diagrammatic sketch of a locomotive type of boiler. Water indicated by dotted lines. The arrows show the direction taken by the air and hot gases from the air-door to the funnel.

THE MULTITUBULAR BOILER.

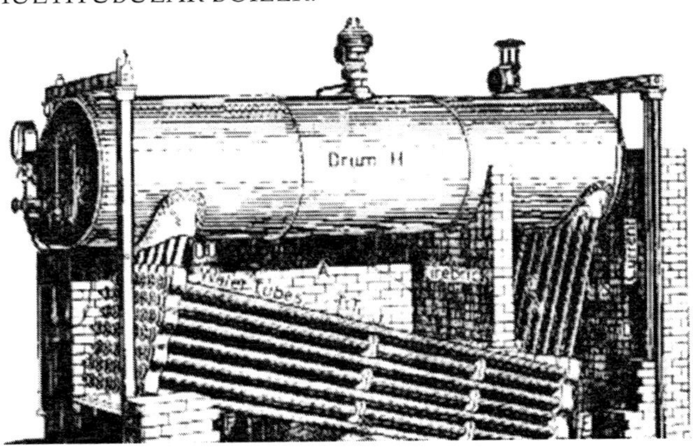

Fig. 7.—The Babcock and Wilcox water-tube boiler. One side of the brick seating has been removed to show the arrangement of the water-tubes and furnace.

To save room, boilers which have to make steam very quickly and at high pressures are largely composed of pipes. Such boilers we call multitubular.

They are of two kinds—(1) *Water*-tube boilers; in which the water circulates through tubes exposed to the furnace heat. The Babcock and Wilcox boiler (Fig. 7) is typical of this variety. (2) *Fire*-tube boilers; in which the hot gases pass through tubes surrounded by water. The ordinary locomotive boiler (Fig. 6) illustrates this form.

The Babcock and Wilcox boiler is widely used in mines, power stations, and, in a modified form, on shipboard. It consists of two main parts—(1) A drum,H, in the upper part of which the steam collects; (2) a group of pipes arranged on the principle illustrated by Fig. 5. The boiler is seated on a rectangular frame of fire-bricks. At one end is the furnace door; at the other the exit to the chimney. From the furnaceFthe flames and hot gases rise round the upper end of the sloping tubesTTinto the spaceA, where they play upon the under surface ofHbefore plunging downward again among the tubes into the spaceB. Here the temperature is lower. The arrows indicate further journeys upwards into the spaceCon the right of a fire-brick division, and past the down tubesSSintoD, whence the hot gases find an escape into the chimney through the openingE. It will be noticed that the greatest heat is brought to bear onTTnear their junction withUU, the "uptake" tubes; and that every succeeding passage of the pipes brings the gradually cooling gases nearer to the "downtake" tubesSS.

The pipesTTare easily brushed and scraped after the removal of plugs from the "headers" into which the tube ends are expanded.

Other well-known water-tube boilers are the Yarrow, Belleville, Stirling, and Thorneycroft, all used for driving marine engines.

FIRE-TUBE BOILERS.

Fig. 6 shows a locomotive boiler in section. To the right is the fire-box, surrounded on all sides by a water-jacket in direct communication with the barrel of the boiler. The inner shell of the fire-box is often made of copper, which withstands the fierce heat better than steel; the outer, like the rest of the boiler, is of steel plates from ½ to ¾ inch thick. The shells of the jacket are braced together by a large number of rivets,RR; and the top, or crown, is strengthened by heavy longitudinal girders riveted to it, or is braced to the top of the boiler by long bolts. A large number of fire-tubes (only three are shown in the diagram for the sake of simplicity) extend from the fire-box to the smoke-box. The most powerful "mammoth" American locomotives have 350 or more tubes, which, with the fire-box, give 4,000 square feet of surface for the furnace heat to act upon. These tubes are expanded at their ends by a special tool into the tube-plates of the fire-box and boiler front. George Stephenson and his predecessors experienced great difficulty in rendering the tube-end joints quite water-tight, but the invention of the "expander" has removed this trouble.

The *fire-brick arch* shown (Fig. 6) in the fire-box is used to deflect the flames towards the back of the fire-box, so that the hot gases may be retarded somewhat, and their combustion rendered more perfect. It also helps to distribute the heat more evenly over the whole of the inside of the box, and prevents cold air from flying directly from the firing door to the tubes. In some American and Continental locomotives the fire-brick arch is replaced by a "water bridge," which serves the same purpose, while giving additional heating surface.

The water circulation in a locomotive boiler is—upwards at the fire-box end, where the heat is most intense; forward along the surface; downwards at the smoke-box end; backwards along the bottom of the barrel.

OTHER TYPES OF BOILERS.

For small stationary land engines the *vertical* boiler is much used. In Fig. 8 we have three forms of this type—AandBwith cross water-tubes;Cwith vertical fire-tubes. The furnace in every case is surrounded by water, and fed through a door at one side.

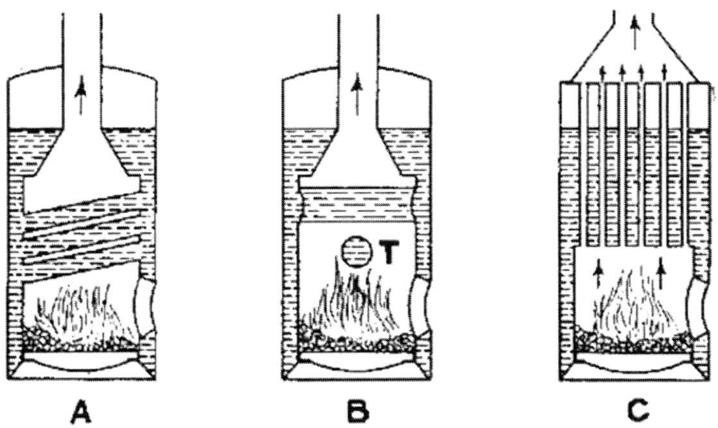

Fig. 8.—Diagrammatic representation of three types of vertical boilers.

The *Lancashire* boiler is of large size. It has a cylindrical shell, measuring up to 30 feet in length and 7 feet in diameter, traversed from end to end by two large flues, in the rear part of which are situated the furnaces. The boiler is fixed on a seating of fire-bricks, so built up as to form three flues,AandBB, shown in cross section in Fig. 9. The furnace gases, after leaving the two furnace flues, are deflected downwards into the channelA, by which they pass underneath the boiler to a point almost under the furnace, where they divide right and left and travel through cross passages into the side channelsBB, to be led along the boiler's flanks to the chimney exitC. By this arrangement the effective heating surface is greatly increased; and the passages being large, natural draught generally suffices to maintain proper combustion. The Lancashire boiler is much used in factories and (in a modified form) on ships, since it is a steady steamer and is easily kept in order.

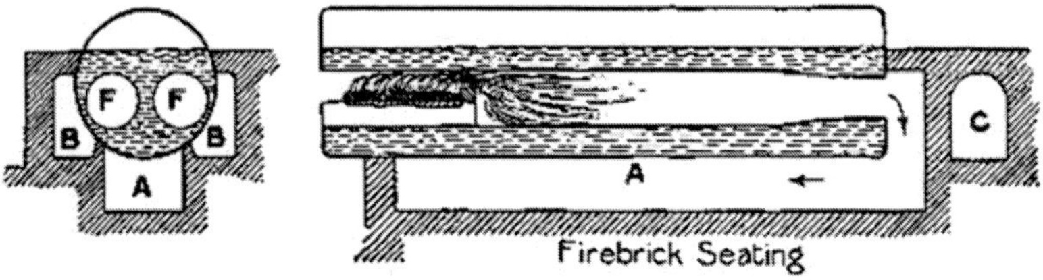

Fig. 9.—Cross and longitudinal sections of a Lancashire boiler.

In marine boilers of cylindrical shape cross water-tubes and fire-tubes are often employed to increase the heating surface. Return tubes are also led through the water to the funnels, situated at the same end as the furnace.

AIDS TO COMBUSTION.

We may now turn our attention more particularly to the chemical process called *combustion*, upon which a boiler depends for its heat. Ordinary steam coal contains about 85 per cent. of carbon, 7 per cent. of oxygen, and 4 per cent. of hydrogen, besides traces of nitrogen and sulphur and a small incombustible residue. When the coal burns, the nitrogen is released and passes away without combining with any of the other elements. The sulphur unites with hydrogen and forms sulphuretted hydrogen (also named sulphurous acid), which is injurious to steel plates, and is largely responsible for the decay of tubes and funnels. More of the hydrogen unites with the oxygen as steam.

The most important element in coal is the carbon (known chemically by the symbol C). Its combination with oxygen, called combustion, is the act which heats the boiler. Only when the carbon present has combined with the greatest possible amount of oxygen that it will take into partnership is the combustion complete and the full heat-value (fixed by scientific experiment at 14,500 thermal units per pound of carbon) developed.

Now, carbon may unite with oxygen, atom for atom, and form *carbon monoxide* (CO); or in the proportion of one atom of carbon to *two* of oxygen, and form *carbon dioxide* (CO2). The former gas is combustible—that is, will admit another atom of carbon to the molecule—but the latter is saturated with oxygen, and will not burn, or, to put it otherwise, is the product of *perfect* combustion. A properly designed furnace, supplied with a due amount of air, will cause nearly all the carbon in the coal burnt to combine with the full amount of oxygen. On the other hand, if the oxygen supply is inefficient, CO as well as CO2 will form, and there will be a heat loss, equal in extreme cases to two-thirds of the whole. It is therefore necessary that a furnace which has to eat up fuel at a great pace should be artificially fed with air in the proportion of from 12 to 20 *pounds* of air for every pound of fuel. There are two methods of creating a violent draught through the furnace. The first is—

The *forced draught*; very simply exemplified by the ordinary bellows used in every house. On a ship (Fig. 10) the principle is developed as follows:—The boilers are situated in a compartment or compartments having no communication with the outer air, except for the passages down which air is forced by powerful fans at a pressure considerably greater than that of the atmosphere. There is only one "way out"—namely, through the furnace and tubes (or gas-ways) of the boiler, and the funnel. So through these it rushes, raising the fuel to white heat. As may easily be imagined, the temperature of a stokehold, especially in the tropics, is far from pleasant. In the Red Sea the thermometer sometimes rises to 170° Fahrenheit or more, and the poor stokers have a very bad time of it.

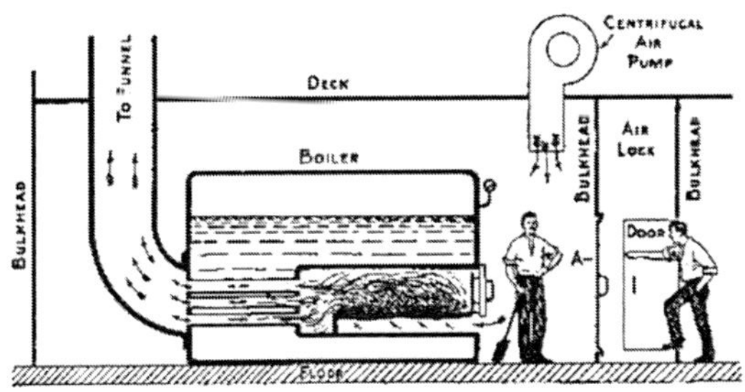

Fig. 10.—Sketch showing how the "forced draught" is produced in a stokehold and how it affects the furnaces.

SCENE IN THE STOKEHOLD OF A BATTLE-SHIP.

The second system is that of the *induced draught*. Here air is *sucked* through the furnace by creating a vacuum in the funnel and in a chamber opening into it. Turning to Fig. 6, we see a pipe through which the exhaust steam from the locomotive's cylinders is shot upwards into the funnel, in which, and in the smoke-box beneath it, a strong vacuum is formed while the engine is running. Now, "nature abhors a vacuum," so air will get into the smoke-box if there be a way open. There is—through the air-doors at the bottom of the furnace, the furnace itself, and the fire-tubes; and on the way oxygen combines with the carbon of the fuel, to form carbon dioxide. The power of the draught is so great that, as one often notices when a train passes during the night, red-hot cinders, plucked from the fire-box, and dragged through the tubes, are hurled far into the air. It might be mentioned in parenthesis that the so-called "smoke" which pours from the funnel of a moving engine is mainly condensing steam. A steamship, on the other hand, belches smoke only from its funnels, as fresh water is far too precious to waste as steam. We shall refer to this later .

BOILER FITTINGS.

The most important fittings on a boiler are:—(1) the safety-valve; (2) the water-gauge; (3) the steam-gauge; (4) the mechanisms for feeding it with water.

THE SAFETY-VALVE.

Professor Thurston, an eminent authority on the steam-engine, has estimated that a plain cylindrical boiler carrying 100 lbs. pressure to the square inch contains sufficient stored energy to project it into the air a vertical distance of 3½ miles. In the case of a Lancashire boiler at equal pressure the distance would be 2½ miles; of a locomotive boiler, at 125 lbs., 1½ miles; of a steam tubular boiler, at 75 lbs., 1 mile. According to the same writer, a cubic foot of heated water under a pressure of from 60 to 70 lbs. per square inch has *about the same energy as one pound of gunpowder.*

Steam is a good servant, but a terrible master. It must be kept under strict control. However strong a boiler may be, it will burst if the steam pressure in it be raised to a certain point; and some device must therefore be fitted on it which will give the steam free egress before that point is reached. A device of this kind is called a *safety-valve*. It usually blows off at less than half the greatest pressure that the boiler has been proved by experiment to be capable of withstanding.

In principle the safety-valve denotes an orifice closed by an accurately-fitting plug, which is pressed against its seat on the boiler top by a weighted lever, or by a spring. As soon as the steam pressure on the face of the plug exceeds the counteracting force of the weight or spring, the plug rises, and steam escapes until equilibrium of the opposing forces is restored.

On stationary engines a lever safety-valve is commonly employed (Fig. 11). The blowing-off point can be varied by shifting the weight along the arm so as to give it a greater or less leverage. On locomotive and marine boilers, where shocks and movements have to be reckoned with, weights are replaced by springs, set to a certain tension, and locked up so that they cannot be tampered with.

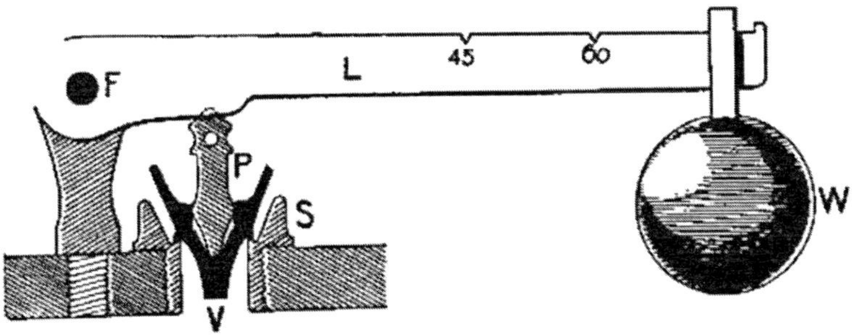

Fig. 11.—A Lever Safety-Valve.V, valve;S, seating;P, pin;L, lever;F, fulcrum;W, weight. The figures indicate the positions at which the weight should be placed for the valve to act when the pressure rises to that number of pounds per square inch.

Boilers are tested by filling the boilers quite full and (1) by heating the water, which expands slightly, but with great pressure; (2) by forcing in additional water with a powerful pump. In either case a rupture would not be attended by an explosion, as water is very inelastic.

The days when an engineer could "sit on the valves"—that is, screw them down—to obtain greater pressure, are now past, and with them a considerable proportion of the dangers of high-pressure steam. The Factory Act of 1895, in force throughout the British Isles, provides that every boiler for generating steam in a factory or workshop where the Act applies must have a proper safety-valve, steam-gauge, and water-gauge; and that boilers and fittings must be examined by a competent person at least once in every fourteen months. Neglect of these provisions renders the owner of a boiler liable to heavy penalties if an explosion occurs.

One of the most disastrous explosions on record took place at the Redcar Iron Works, Yorkshire, in June 1895. In this case, twelve out of fifteen boilers ranged side by side burst, through one proving too weak for its work. The flying fragments of this boiler, striking the sides of other boilers, exploded them, and so the damage was transmitted down the line. Twenty men were killed and injured; while masses of metal, weighing several tons each, were hurled 250 yards, and caused widespread damage.

The following is taken from a journal, dated December 22, 1895: " *Providence* (*Rhode Island*).—A recent prophecy that a boiler would explode between December 16 and 24 in a store has seriously affected the Christmas trade. Shoppers are incredibly nervous. One store advertises, 'No boilers are being used; lifts running electrically.' All stores have had their boilers inspected."

THE WATER-GAUGE.

No fitting of a boiler is more important than the *water-gauge,* which shows the level at which the water stands. The engineer must continually consult his gauge, for if the water gets too low, pipes and other surfaces exposed to the furnace flames may burn through, with disastrous results; while, on the other hand, too much water will cause bad steaming. A section of an ordinary gauge is seen in Fig. 12. It consists of two parts, each furnished with a gland,G, to make a steam-tight joint round the glass tube, which is inserted through the hole covered by the plugP1. The cocksT1 T2 are normally open, allowing the ingress of steam and water respectively to the tube. CockT3 is kept closed unless for any reason it is necessary to blow steam or water through the gauge. The holesC Ccan be cleaned out if the plugsP2 P3 are removed.

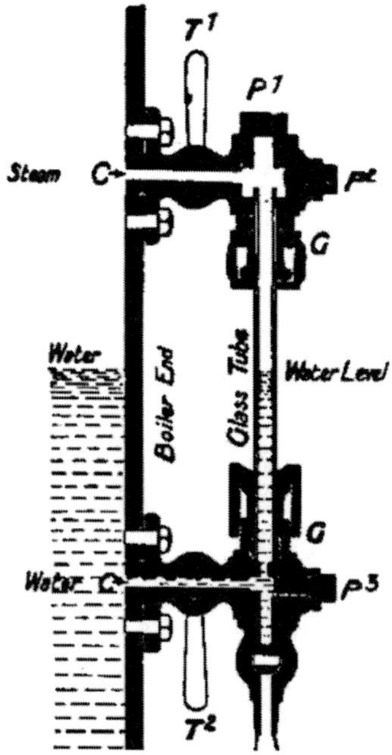

Fig. 12.—Section of a water-gauge.

Most gauges on high-pressure boilers have a thick glass screen in front, so that in the event of the tube breaking, the steam and water may not blow directly on to the attendants. A further precaution is to include two ball-valves near the ends of the gauge-glass. Under ordinary conditions the balls lie in depressions clear of the ways; but when a rush of steam or water occurs they are sucked into their seatings and block all egress.

On many boilers two water-gauges are fitted, since any gauge may work badly at times. The glasses are tested to a pressure of 3,000 lbs. or more to the square inch before use.

THE STEAM-GAUGE.

It is of the utmost importance that a person in charge of a boiler should know what pressure the steam has reached. Every boiler is therefore fitted with one *steam-gauge*; many with two, lest one might be unreliable. There are two principal types of steam-gauge:—(1) The Bourdon; (2) the Schäffer-Budenberg. The principle of the Bourdon is illustrated by Fig. 13, in whichAis a piece of rubber tubing closed at one end, and at the other drawn over the nozzle of a cycle tyre inflatoı. If bent in a curve, as shown, the section of the tube is an oval. When air is pumped in, the rubber walls endeavour to assume a circular section, because this shape encloses a larger area than an oval of equal circumference, and therefore makes room for a larger volume of air. In doing so the tube straightens itself, and assumes the position indicated by the dotted lines. Hang an empty "inner tube" of a pneumatic tyre over a nail and inflate it, and you will get a good illustration of the principle.

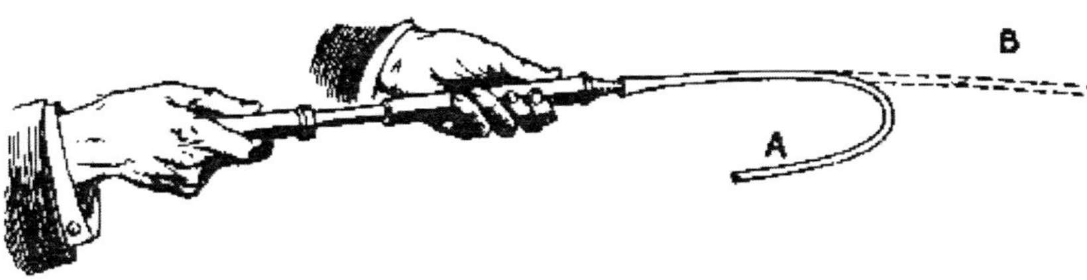

Fig. 13.—Showing the principle of the steam-gauge.

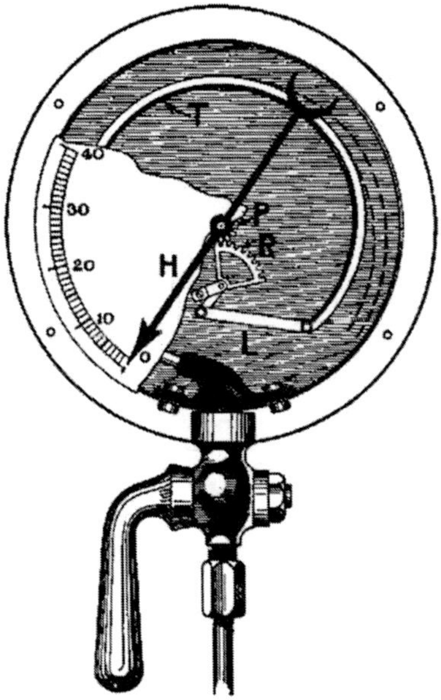

Fig. 14.—Bourdon steam-gauge. Part of dial removed to show mechanism.

In Fig. 14 we have a Bourdon gauge, with part of the dial face broken away to show the internal mechanism.Tis a flattened metal tube soldered at one end into a hollow casting, into which screws a tap connected with the boiler. The other end (closed) is attached to a link,L, which works an arm of a quadrant rack,R, engaging with a small pinion,P, actuating the pointer. As the steam pressure rises, the tubeTmoves its free end outwards towards the position shown by the dotted lines, and traverses the arm of the rack, so shifting the pointer round the scale. As the pressure falls, the tube gradually returns to its zero position.

The Schäffer-Budenberg gauge depends for its action on the elasticity of a thin corrugated metal plate, on one side of which steam presses. As the plate bulges upwards it pushes up a small rod resting on it, which operates a quadrant and rack similar to that of the Bourdon gauge. The principle is employed in another form for the aneroid barometer.

THE WATER SUPPLY TO A BOILER.

The water inside a boiler is kept at a proper level by (1) pumps or (2) injectors. The former are most commonly used on stationary and marine boilers. As their mechanism is much the same as that of ordinary force pumps, which will be described in a later chapter, we may pass at once to the *injector*, now almost universally used on locomotive, and sometimes on stationary boilers. At first sight the injector is a mechanical paradox, since it employs the steam from a boiler to blow water into the boiler. In Fig. 15 we have an illustration of the principle of an injector. Steam is led from the boiler through pipeA, which terminates in a nozzle surrounded by a cone,E, connected by the pipeBwith the water tank. When steam is turned on it rushes with immense velocity from the nozzle, and creates a partial vacuum in coneE, which soon fills with water. On meeting the water the steam condenses, but not before it has imparted some of its *velocity* to the water, which thus gains sufficient momentum to force down the valve and find its way to the boiler. The overflow spaceO ObetweenEandCallows steam and water to escape until the water has gathered the requisite momentum.

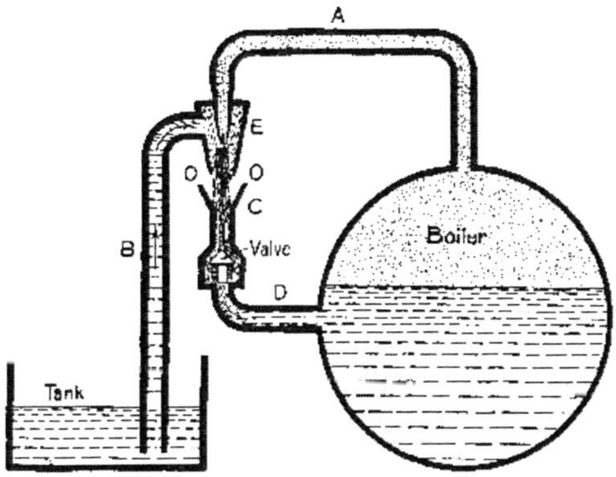

Fig. 15.—Diagram illustrating the principle of a steam-injector.

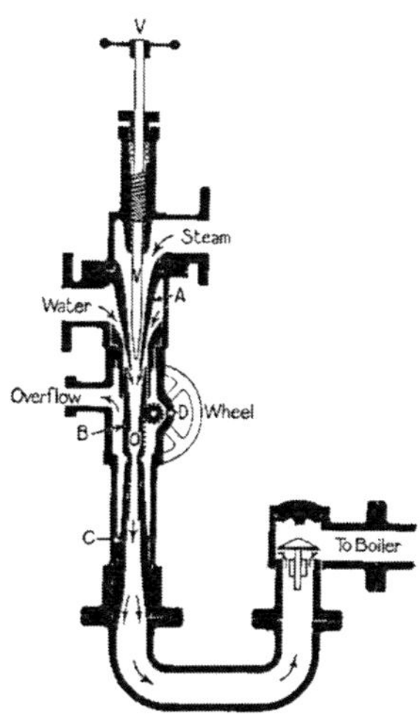

Fig. 16.—The Giffard injector.

A form of injector very commonly used is Giffard's (Fig. 16). Steam is allowed to enter by screwing up the valveV. As it rushes through the nozzle of the coneAit takes up water and projects it into the "mixing cone"B, which can be raised or lowered by the pinionD(worked by the hand-wheel wheel shown) so as to regulate the amount of water admitted toB. At the centre ofBis an aperture,O, communicating with the overflow. The water passes to the boiler through the valve on the left. It will be noticed that the coneAand the part ofBabove the orificeOcontract downward. This is to convert the *pressure* of the steam into *velocity*. BelowOis a cone, the diameter of which increases downwards. Here the *velocity* of the water is converted back into *pressure* in obedience to a well-known hydromechanic law.

An injector does not work well if the feed-water be too hot to condense the steam quickly; and it may be taken as a rule that the warmer the water, the smaller is the amount of it injected by a given weight of steam.[2] Some injectors have flap-valves covering the overflow orifice, to prevent air being sucked in and carried to the boiler.

When an injector receives a sudden shock, such as that produced by the passing of a locomotive over points, it is liable to "fly off"—that is, stop momentarily—and then send the steam and water through the overflow. If this happens, both steam and water must be turned off, and the injector be restarted; unless it be of the *self-starting* variety, which automatically controls the admission of water to the "mixing-cone," and allows the injector to "pick up" of itself.

For economy's sake part of the steam expelled from the cylinders of a locomotive is sometimes used to work an injector, which passes the water on, at a pressure of 70 lbs. to the square inch, to a second injector operated by high-pressure steam coming direct from the boiler, which increases its velocity sufficiently to overcome the boiler pressure. In this case only a fraction of the weight of high-pressure steam is required to inject a given weight of water, as compared with that used in a single-stage injector.

[1] "The Steam-Engine,".

[2] By "weight of steam" is meant the steam produced by boiling a certain weight of water. A pound of steam, if condensed, would form a pound of water.

CHAPTER II.
THE CONVERSION OF HEAT ENERGY INTO MECHANICAL MOTION.

Reciprocating engines—Double-cylinder engines—The function of the fly-wheel—The cylinder—The slide-valve—The eccentric—"Lap" of the valve: expansion of steam—How the cut-off is managed—Limit of expansive working—Compound engines—Arrangement of expansion engines—Compound locomotives—Reversing gears—"Linking-up"—Piston-valves—Speed governors—Marine-speed governors—The condenser.

HAVING treated at some length the apparatus used for converting water into high-pressure steam, we may pass at once to a consideration of the mechanisms which convert the energy of steam into mechanical motion, or *work*.

Steam-engines are of two kinds:—(1) *reciprocating*, employing cylinders and cranks; (2) *rotary*, called turbines.

RECIPROCATING ENGINES.

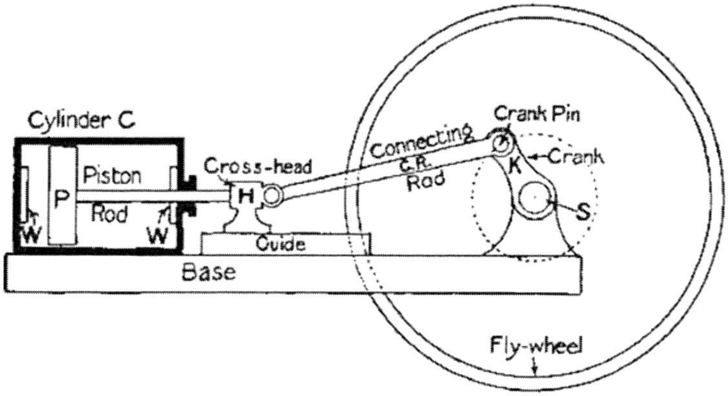

Fig. 17.—Sketch showing parts of a horizontal steam-engine.

Fig. 17 is a skeleton diagram of the simplest form of reciprocating engine. Cis a *cylinder* to which steam is admitted through the *steam-ways* [3]W W, first on one side of the pistonP, then on the other. The pressure on the piston pushes it along the cylinder, and the force is transmitted through the piston rodP Rto the *connecting rod*C R, which causes the *crank*Kto revolve. At the point where the two rods meet there is a "crosshead,"H, running to and fro in a guide to prevent the piston rod being broken or bent by the oblique thrusts and pulls which it imparts throughC Rto the crankK. The latter is keyed to a *shaft*Scarrying the fly-wheel, or, in the case of a locomotive, the driving-wheels. The crank shaft revolves in bearings. The internal diameter of a cylinder is called its *bore*. The travel of the piston is called its *stroke*. The distance from the centre of the shaft to the centre of the crank pin is called the crank's *throw,* which is half of the piston's *stroke*. An engine of this type is called double-acting, as the piston is pushed alternately backwards and forwards by the steam. When piston rod, connecting rod, and crank lie in a straight line—that is, when the piston is fully out, or fully in—the crank is said to be at a "dead point;" for, were the crank turned to such a position, the admission of steam would not produce motion, since the thrust or pull would be entirely absorbed by the bearings.

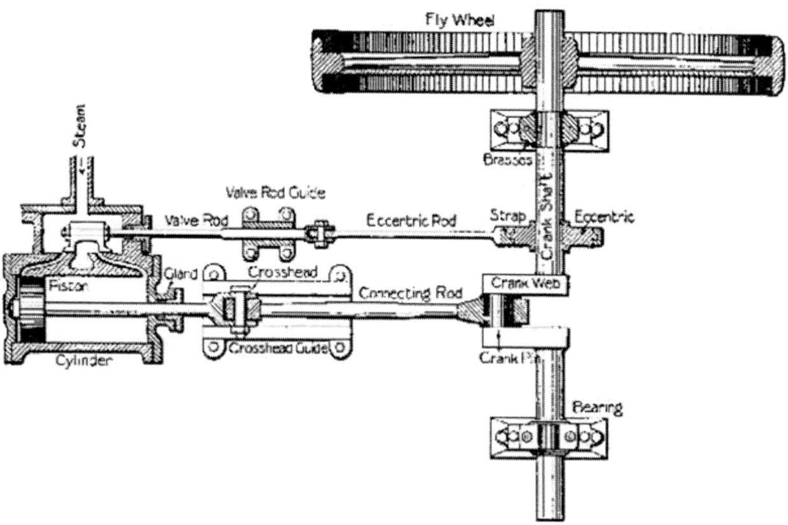

Fig. 18.—Sectional plan of a horizontal engine.

DOUBLE-CYLINDER ENGINES.

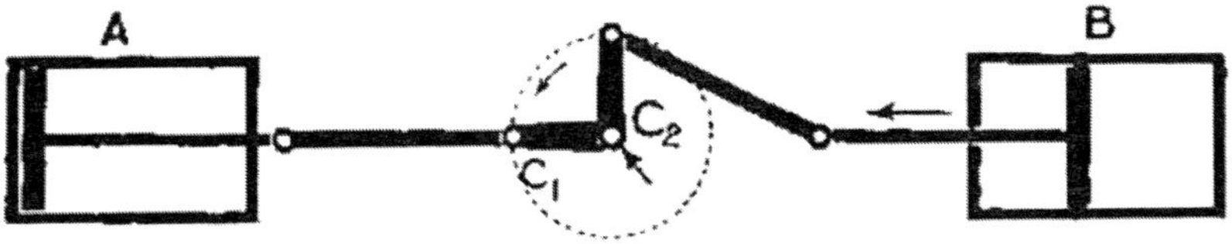

Fig. 19.

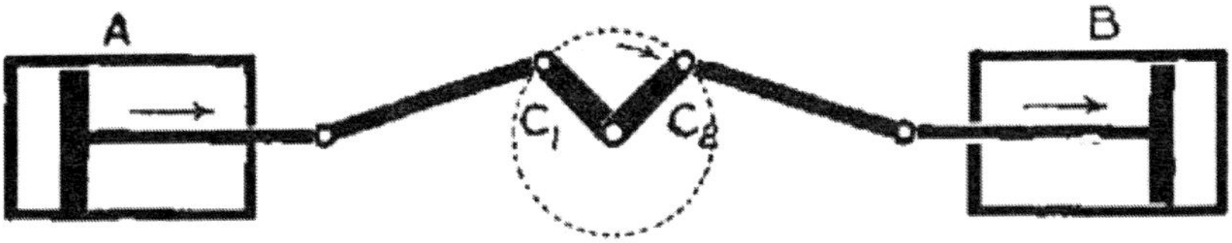

Fig. 20.

Locomotive, marine, and all other engines which must be started in any position have at least *two* cylinders, and as many cranks set at an angle to one another. Fig. 19 demonstrates that when one crank,C1, of a double-cylinder engine is at a "dead point," the other,C2, has reached a position at which the piston exerts the maximum of turning power. In Fig. 20 each crank is at 45° with the horizontal, and both pistons are able to do work. The power of one piston is constantly increasing while that of the other is decreasing. If *single*-action cylinders are used, at least *three* of these are needed to produce a perpetual turning movement, independently of a fly-wheel.

THE FUNCTION OF THE FLY-WHEEL.

A fly-wheel acts as a *reservoir of energy*, to carry the crank of a single-cylinder engine past the "dead points." It is useful in all reciprocating engines to produce steady running, as a heavy wheel acts as a drag on the effects of a sudden increase or decrease of steam pressure. In a pump, mangold-slicer, cake-crusher, or chaff-cutter, the fly-wheel helps the operator to pass *his* dead points—that is, those parts of the circle described by the handle in which he can do little work.

THE CYLINDER.

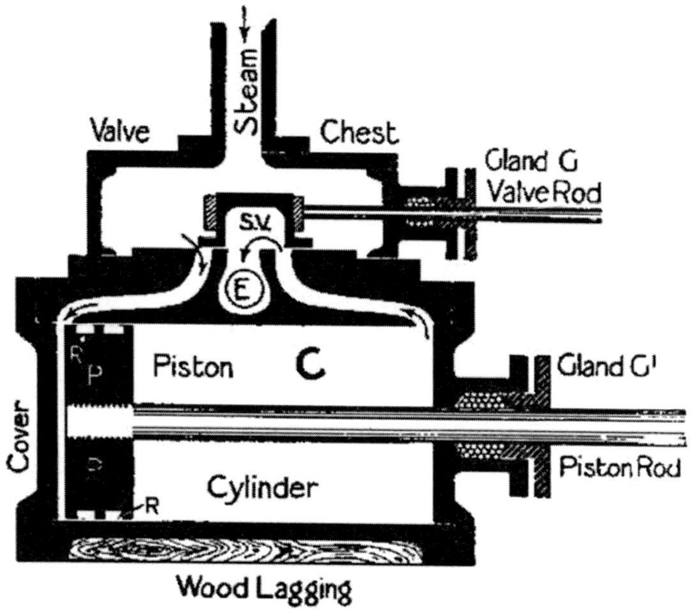

Fig. 21.—Diagrammatic section of a cylinder and its slide-valve.

The cylinders of an engine take the place of the muscular system of the human body. In Fig. 21 we have a cylinder and its slide-valve shown in section. First of all, look atP, the piston. Round it are white grooves,R R, in which rings are fitted to prevent the passage of steam past the piston. The rings are cut through at one point in their circumference, and slightly opened, so that when in position they press all round against the walls of the cylinder. After a little use they "settle down to their work"—that is, wear to a true fit in the cylinder. Each end of the cylinder is closed by a cover, one of which has a boss cast on it, pierced by a hole for the piston rod to work through. To prevent the escape of steam the boss is hollowed out true to accommodate a *gland*,G1, which is threaded on the rod and screwed up against the boss; the internal space between them being filled with packing. Steam from the boiler enters the steam-chest, and would have access to both sides of the piston simultaneously through the steam-ways,W W, were it not for the

SLIDE-VALVE,

a hollow box open at the bottom, and long enough for its edges to cover both steam-ways at once. BetweenW WisE, the passage for the exhaust steam to escape by. The edges of the slide-valve are perfectly flat, as is the face over which the valve moves, so that no steam may pass under the edges. In our illustration the piston has just begun to move towards the right. Steam enters by the left steam-way, which the valve is just commencing to uncover. As the piston moves, the valve moves in the same direction until the port is fully uncovered, when it begins to move back again; and just before the piston has finished its stroke the steam-way on the right begins to open. The steam-way on the left is now in communication with the exhaust portE, so that the steam that has done its duty is released and pressed from the cylinder by the piston. *Reciprocation* is this backward and forward motion of the piston: hence the term "reciprocating" engines. The linear motion of the piston rod is converted into rotatory motion by the connecting rod and crank.

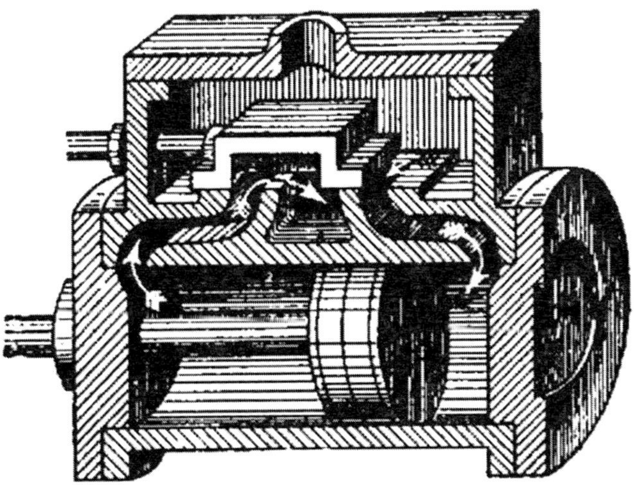

Fig. 22.—Perspective section of cylinder.

The use of a crank appears to be so obvious a method of producing this conversion that it is interesting to learn that, when James Watt produced his "rotative engine" in 1780 he was unable to use the crank because it had already been patented by one Matthew Wasborough. Watt was not easily daunted, however, and within a twelvemonth had himself patented five other devices for obtaining rotatory motion from a piston rod. Before passing on, it may be mentioned that Watt was the father of the modern—that is, the high-pressure—steam-engine; and that, owing to the imperfection of the existing machinery, the difficulties he had to overcome were enormous. On one occasion he congratulated himself because one of his steam-cylinders was only three-eighths of an inch out of truth in the bore. Nowadays a good firm would reject a cylinder1⁄500 of an inch out of truth; and in small petrol-engines1⁄5000 of an inch is sometimes the greatest "limit of error" allowed.

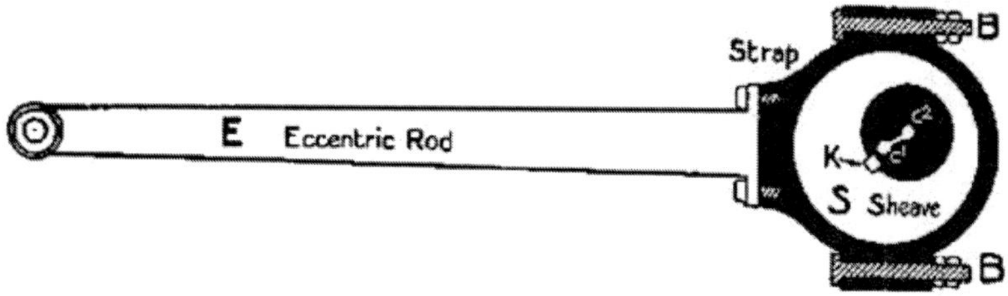

Fig. 23.—The eccentric and its rod.

THE ECCENTRIC

is used to move the slide-valve to and fro over the steam ports (Fig. 23). It consists of three main parts—the *sheave*, or circular plateS, mounted on the crank shaft; and the two *straps* which encircle it, and in which it revolves. To one strap is bolted the "big end" of the eccentric rod, which engages at its other end with the valve rod. The straps are semicircular and held together by strong bolts,B B, passing through lugs, or thickenings at the ends of the semicircles. The sheave has a deep groove all round the edges, in which the straps ride. The "eccentricity" or "throw" of an eccentric is the distance betweenC2, the centre of the shaft, andC1, the centre of the sheave. The throw must equal half of the distance which the slide-valve has to travel over the steam ports. A tapering steel wedge or key,K, sunk half in the eccentric and half in a slot in the shaft, holds the eccentric steady and prevents it slipping. Some eccentric sheaves are made in two parts, bolted together, so that they may be removed easily without dismounting the shaft.

The eccentric is in principle nothing more than a crank pin so exaggerated as to be larger than the shaft of the crank. Its convenience lies in the fact that it may be mounted at any point on a shaft, whereas a crank can be situated at an end only, if it is not actually aV-shaped bend in the shaft itself—in which case its position is of course permanent.

SETTING OF THE SLIDE-VALVE AND ECCENTRIC.

The subject of valve-setting is so extensive that a full exposition might weary the reader, even if space permitted its inclusion. But inasmuch as the effectiveness of a reciprocating engine depends largely on the nature and arrangement of the valves, we will glance at some of the more elementary principles.

Fig. 24.

Fig. 25.

In Fig. 24 we see in section the slide-valve, the ports of the cylinder, and part of the piston. To the right are two lines at right angles—the thicker,C, representing the position of thc crank; the thinner,E, that of the eccentric. (The position of an eccentric is denoted diagrammatically by a line drawn from the centre of the crank shaft through the centre of the sheave.) The edges of the valve are in this case only broad enough to just cover the ports—that is, they have no *lap*. The piston is about to commence its stroke towards the left; and the eccentric, which is set at an angle of 90° in *advance* of the crank, is about to begin opening the left-hand port. By the time thatChas got to the position originally occupied byE,Ewill be horizontal (Fig. 25)—that is, the eccentric will have finished its stroke towards the left; and whileCpasses through the next right angle the valve will be closing the left port, which will cease to admit steam when the piston has come to the end of its travel. The operation is repeated on the right-hand side while the piston returns.

Fig. 26.

It must be noticed here—(1) that steam is admitted at full pressure *all through* the stroke; (2) that admission begins and ends simultaneously with the stroke. Now, in actual practice it is necessary to admit steam before the piston has ended its travel, so as to *cushion* the violence of the sudden change of direction of the piston, its rod, and other moving parts. To effect this, the eccentric is set more than 90° in advance—that is, more than what the engineers call *square*. Fig. 26 shows such an arrangement. The angle betweenEandE1 is called the *angle of advance*. Referring to the valve, you will see that it has opened an appreciable amount, though the piston has not yet started on its rightwards journey.

"LAP" OF THE VALVE—EXPANSION OF STEAM.

In the simple form of valve that appears in Fig. 24, the valve faces are just wide enough to cover the steam ports. If the eccentric is not *square* with the crank, the admission of steam lasts until the very end of the stroke; if set a little in advance—that is, given *lead*—the steam is cut off before the piston has travelled quite along the cylinder, and readmitted before the back stroke is accomplished. Even with this lead the working is very uneconomical, as the steam goes to the exhaust at practically the same pressure as that at which it entered the cylinder. Its property of *expansion* has been neglected. But supposing that steam at 100 lbs. pressure were admitted till half-stroke, and then suddenly cut off, the expansive nature of the steam would then continue to push the piston out until the pressure had decreased to 50 lbs. per square inch, at which pressure it would go to the exhaust. Now, observe that all the work done by the steam after the cut-off is so much power saved. The *average* pressure on the piston is not so high as in the first case; still, from a given volume of 100 lbs. pressure steam we get much more *work*.

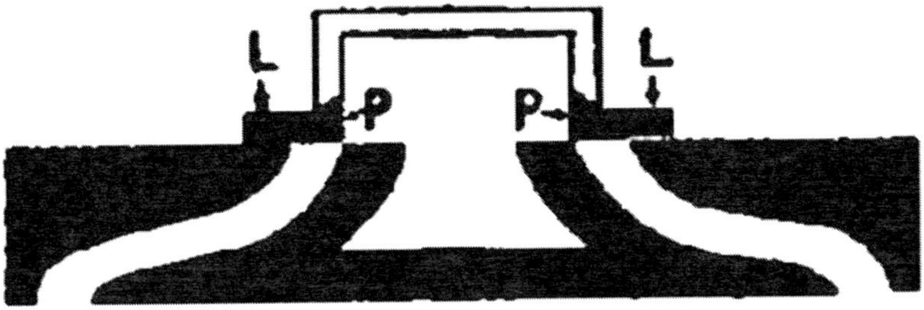

Fig. 27.—A slide-valve with "lap."

Fig. 28.

Look at Fig. 27. Here we have a slide-valve, with faces much wider than the steam ports. The parts marked black,P P, are those corresponding to the faces of the valves shown in previous diagrams. The shaded parts,L L, are called the *lap*. By increasing the length of the lap we increase the range of expansive working. Fig. 28 shows the piston full to the left; the valve is just on the point of opening to admit steam behind the piston. The eccentric has a throw equal to the breadth of a port the lap of the valve. That this must be so is obvious from a consideration of Fig. 27, where the valve is at its central position. Hence the very simple formula:—Travel of valve = 2 × (lap breadth of port). The path of the eccentric's centre round the centre of the shaft is indicated by the usual dotted line (Fig. 28). You will notice that the "angle of advance," denoted by the arrowA, is now very considerable. By the time that the crankChas assumed the position of the lineS, the eccentric has passed its dead point, and the valve begins to travel backwards, eventually returning to the position shown in Fig. 28, and cutting off the steam supply while the piston has still a considerable part of its stroke to make. The steam then begins to work expansively, and continues to do so until the valve assumes the position shown in Fig. 27.

If the valve has to have "lead" to admit steam *before* the end of the stroke to the other side of the piston, the *angle of advance* must be increased, and the eccentric centre line would lie on the lineE2. Therefore—total angle of advance = angle for *lap* and angle for *lead*.

LIMIT OF EXPANSIVE WORKING.

Theoretically, by increasing the *lap* and cutting off the steam earlier and earlier in the stroke, we should economize our power more and more. But in practice a great difficulty is met with—namely, that *as the steam expands its temperature falls*. If the cut-off occurs early, say at one-third stroke, the great expansion will reduce the temperature of the metal walls of the cylinder to such an extent, that when the next spirt of steam enters from the other end a considerable proportion of the steam's energy will be lost by cooling. In such a case, the difference in temperature between admitted steam and exhausted steam is too great for economy. Yet we want to utilize as much energy as possible. How are we to do it?

COMPOUND ENGINES.

In the year 1853, John Elder, founder of the shipping firm of Elder and Co., Glasgow, introduced the *compound* engine for use on ships. The steam, when exhausted from the high-pressure cylinder, passed into another cylinder of equal stroke but larger diameter, where the expansion continued. In modern engines the expansion is extended to three and even four stages, according to the boiler pressure; for it is a rule that the higher the initial pressure is, the larger is the number of stages of expansion consistent with economical working.

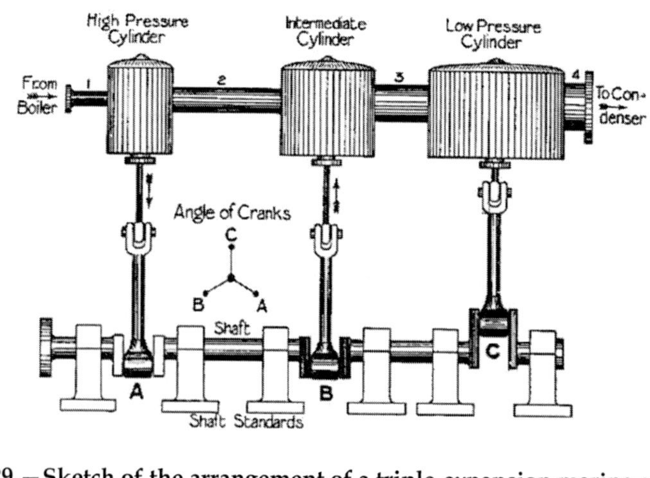

Fig. 29.—Sketch of the arrangement of a triple-expansion marine engine. No valve gear or supports, etc., shown.

In Fig. 29 we have a triple-expansion marine engine. Steam enters the high-pressure cylinder[4] at, say, 200 lbs. per square inch. It exhausts at 75 lbs. into the large pipe 2, and passes to the intermediate cylinder, whence it is exhausted at 25 lbs. or so through pipe 3 to the low-pressure cylinder. Finally, it is ejected at about 8 lbs. per square inch to the condenser, and is suddenly converted into water; an act which produces a vacuum, and diminishes the back-pressure of the exhaust from cylinderC. In fact, the condenser exerts a *sucking* power on the exhaust side ofC's piston.

ARRANGEMENT OF EXPANSION ENGINES.

In the illustration the cranks are set at angles of 120°, or a third of a circle, so that one or other is always at or near the position of maximum turning power. Where only two stages are used the cylinders are often arranged *tandem*, both pistons having a common piston rod and crank. In order to get a constant turning movement they must be mounted separately, and work cranks set at right angles to one another.

COMPOUND LOCOMOTIVES.

In 1876 Mr. A. Mallet introduced *compounding* in locomotives; and the practice has been largely adopted. The various types of "compounds" may be classified as follows:—(1) One low-pressure and one high-pressure cylinder; (2) one high-pressure and two low-pressure; (3) one low-pressure and two high-pressure; (4) two high-pressure and two low-pressure. The last class is very widely used in France, America, and Russia, and seems to give the best results. Where only two cylinders are used (and sometimes in the case of three and four), a valve arrangement permits the admission of high-pressure steam to both high and low-pressure cylinders for starting a train, or moving it up heavy grades.

REVERSING GEARS.

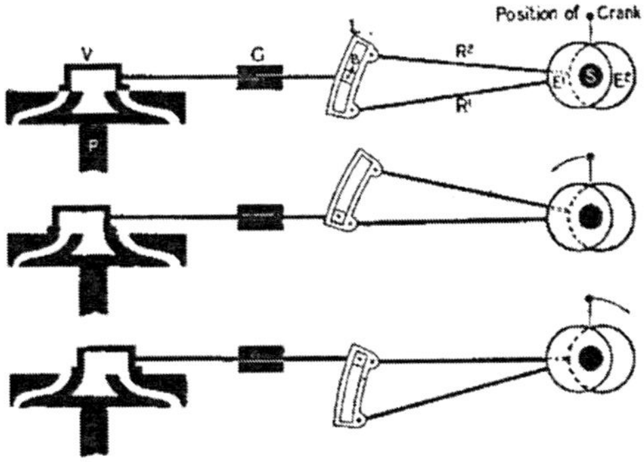

Figs. 30, 31, 32.—Showing how a reversing gear alters the position of the slide-valve.

The engines of a locomotive or steamship must be reversible—that is, when steam is admitted to the cylinders, the engineer must be able to so direct it through the steam-ways that the cranks may turn in the desired direction. The commonest form of reversing device (invented by George Stephenson) is known as Stephenson's Link Gear. In Fig. 30 we have a diagrammatic presentment of this gear.E1 andE2 are two eccentrics set square with the crank at opposite ends of a diameter. Their rods are connected to the ends of a link,L, which can be raised and lowered by means of levers (not shown). Bis a block which can partly revolve on a pin projecting from the valve rod, working through a guide,G. In Fig. 31 the link is half raised, or in "mid-gear," as drivers say. EccentricE1 has pushed the lower end of the link fully back;E2 has pulled it fully forward; and since any movement of the one eccentric is counterbalanced by the opposite movement of the other, rotation of the eccentrics would not cause the valve to move at all, and no steam could be admitted to the cylinder.

Let us suppose that Fig. 30 denotes one cylinder, crank, rods, etc., of a locomotive. The crank has come to rest at its half-stroke; the reversing lever is at the mid-gear notch. If the engineer desires to turn his cranks in an anti-clockwise direction, he *raises* the link, which brings the rod ofE1 into line with the valve rod and presses the block *backwards* till the right-hand port is uncovered (Fig. 31). If steam be now admitted, the piston will be pushed towards the left, and the engine will continue to run in an anti-clockwise direction. If, on the other hand, he wants to run the engine the other way, he would *drop* the link, bringing the rod ofE2 into line with the valve rod, and drawingV *forward* to uncover the rear port (Fig. 32). In either case the eccentric working the end of the link remote fromBhas no effect, since it merely causes that end to describe arcs of circles of whichBis the centre.

"LINKING UP."

If the link is only partly lowered or raised from the central position it still causes the engine to run accordingly, but the movement of the valve is decreased. When running at high speed the engineer "links up" his reversing gear, causing his valves to cut off early in the stroke, and the steam to work more expansively than it could with the lever at *full*, or *end*, gear; so that this device not only renders an engine reversible, but also gives the engineer an absolute command over the expansion ratio of the steam admitted to the cylinder, and furnishes a method of cutting off the steam altogether. In Figs. 30, 31, 32, the valve has no lap and the eccentrics are set square. In actual practice the valve faces would have "lap" and the eccentric "lead" to correspond; but for the sake of simplicity neither is shown.

OTHER GEARS.

In the Gooch gear for reversing locomotives the link does not shift, but the valve rod and its block is raised or lowered. The Allan gear is so arranged that when the link is raised the block is lowered, and *vice versâ*. These are really only modifications of Stephenson's principle—namely, the employment of *two* eccentrics set at equal angles to and on opposite sides of the crank. There are three other forms of link-reversing gear, and nearly a dozen types of *radial* reversing devices; but as we have already described the three most commonly used on locomotives and ships, there is no need to give particulars of these.

Before the introduction of Stephenson's gear a single eccentric was used for each cylinder, and to reverse the engine this eccentric had to be loose on the axle. "A lever and gear worked by a treadle on the footplate controlled the position of the eccentrics. When starting the engine, the driver put the eccentrics out of gear by the treadle; then, by means of a lever he raised the small-ends[5] of the eccentric rods, and, noting the position of the cranks, or, if more convenient, the balance weight in the wheels, he, by means of another handle, moved the valves to open the necessary ports to steam and worked them by hand until the engine was moving; then, with the treadle, he threw the eccentrics over to engage the studs, at the same time dropping the small-ends of the rods to engage pins upon the valve spindles, so that they continued to keep up the movement of the valve."[6] One would imagine that in modern shunting yards such a device would somewhat delay operations!

PISTON VALVES.

In marine engines, and on many locomotives and some stationary engines, theD-valve (shown in Figs. 30–32) is replaced by a piston valve, or circular valve, working up and down in a tubular seating. It may best be described as a rod carrying two pistons which correspond to the faces of aD-valve. Instead of rectangular ports there are openings in the tube in which the piston valve moves, communicating with the steam-ways into the cylinder and with the exhaust pipe. In the case of theD-valve the pressure above it is much greater than that below, and considerable friction arises if the rubbing faces are not kept well lubricated. The piston valve gets over this difficulty, since such steam as may leak past it presses on its circumference at all points equally.

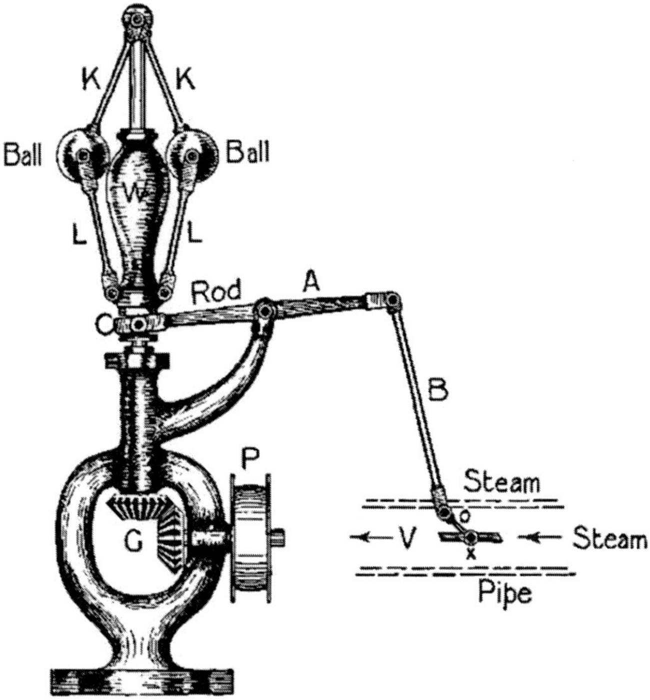

Fig. 33.—A speed governor.

Practically all engines except locomotives and those known as "donkey-engines"—used on cranes—are fitted with some device for keeping the rotatory speed of the crank constant within very narrow limits. Perhaps you have seen a pair of balls moving round on a seating over the boiler of a threshing-engine. They form part of the "governor," or speed-controller, shown in principle in Fig. 33. A belt driven by a pulley on the crank shaft turns a small pulley,P, at the foot of the governor. This transmits motion through two bevel-wheels,G, to a vertical shaft, from the top of which hang two heavy balls on links,K K. Two more links,L L, connect the balls with a weight,W, which has a deep groove cut round it at the bottom. When the shaft revolves, the balls fly outwards by centrifugal force, and as their velocity increases the quadrilateral figure contained by the four links expands laterally and shortens vertically. The angles betweenK KandL Lbecome less and less obtuse, and the weightWis drawn upwards, bringing with it the forkCof the rodA, which has ends engaging with the groove. AsCrises, the other end of the rod is depressed, and the rodBdepresses rodO, which is attached to the spindle operating a sort of shutter in the steam-pipe. Consequently the supply of steam is throttled more and more as the speed increases, until it has been so reduced that the engine slows, and the balls fall, opening the valve again. Fig. 34 shows the valve fully closed. This form of governor was invented by James Watt. A spring is often used instead of a weight, and the governor is arranged horizontally so that it may be driven direct from the crank shaft without the intervention of bevel gearing.

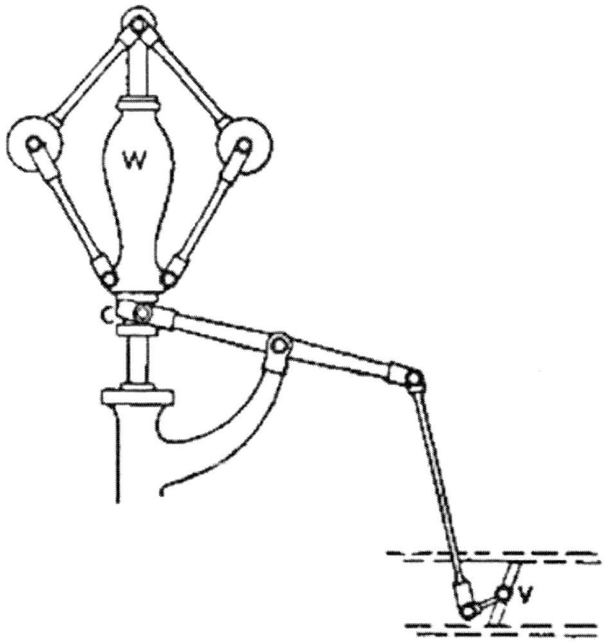

Fig. 34.

The Hartwell governor employs a link motion. You must here picture the balls raising and lowering the *free end* of the valve rod, which carries a block moving in a link connected with the eccentric rod. The link is pivoted at the upper end, and the eccentric rod is attached to the lower. When the engine is at rest the end of the valve rod and its block are dropped till in a line with the eccentric rod; but when the machinery begins to work the block is gradually drawn up by the governor, diminishing the movement of the valve, and so shortening the period of steam admission to the cylinder.

Governors are of special importance where the *load* of an engine is constantly varying, as in the case of a sawmill. A good governor will limit variation of speed within two per cent.—that is, if the engine is set to run at 100 revolutions a minute, it will not allow it to exceed 101 or fall below 99.

In *very* high-speed engines the governing will prevent variation of less than one per cent., even when the load is at one instant full on, and the next taken completely off.

MARINE GOVERNORS.

These must be more quick-acting than those used on engines provided with fly-wheels, which prevent very sudden variations of speed. The screw is light in proportion to the engine power, and when it is suddenly raised from the water by the pitching of the vessel, the engine would race till the screw took the water again, unless some regulating mechanism were provided. Many types of marine governors have been tried. The most successful seems to be one in which water is being constantly forced by a pump driven off the engine shaft into a cylinder controlling a throttle-valve in the main steam-pipe. The water escapes through a leak, which is adjustable. As long as the speed of the engine is normal, the water escapes from the cylinder as fast as it is pumped in, and no movement of the piston results; but when the screw begins to race, the pump overcomes the leak, and the piston is driven out, causing a throttling of the steam supply.

CONDENSERS.

The *condenser* serves two purposes:—(1) It makes it possible to use the same water over and over again in the boilers. On the sea, where fresh water is not obtainable in large quantities, this is a matter of the greatest importance. (2) It adds to the power of a compound engine by exerting a back pull on the piston of the low-pressure cylinder while the steam is being exhausted.

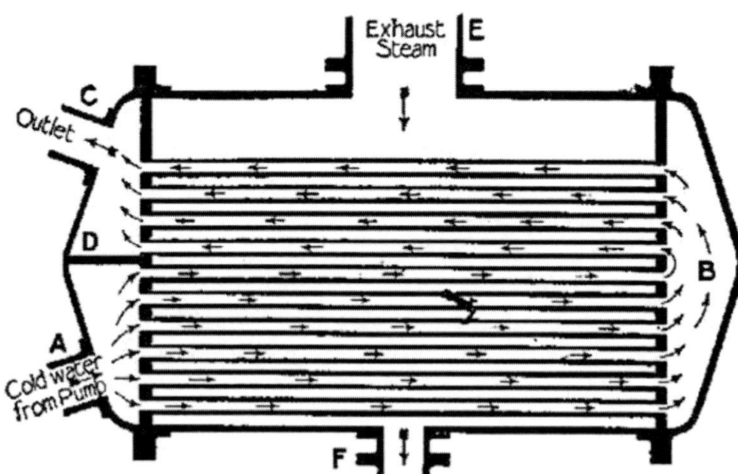

Fig. 35.—The marine condenser.

Fig. 35 is a sectional illustration of a marine condenser. Steam enters the condenser through the large pipeE, and passes among a number of very thin copper tubes, through which sea-water is kept circulating by a pump. The path of the water is shown by the featherless arrows. It comes from the pump through pipeAinto the lower part of a large cap covering one end of the condenser and divided transversely by a diaphragm,D. Passing through the pipes, it reaches the cap attached to the other end, and flows back through the upper tubes to the outletC. This arrangement ensures that, as the steam condenses, it shall meet colder and colder tubes, and finally be turned to water, which passes to the well through the outletF. In some condensers the positions of steam and water are reversed, steam going through the tubes outside which cold water circulates.

[3] Also called *ports.*

[4] The bores of the cylinders are in the proportion of 4: 6: 9. The stroke of all three is the same.

[5] The ends furthest from the eccentric.

[6] "The Locomotive of To-day,".

CHAPTER III.
THE STEAM TURBINE.

How a turbine works—The De Laval turbine—The Parsons turbine—Description of the Parsons turbine—The expansive action of steam in a Parsons turbine—Balancing the thrust—Advantages of the marine turbine.

MORE than two thousand years ago Hero of Alexandria produced the first apparatus to which the name of steam-engine could rightly be given. Its principle was practically the same as that of the revolving jet used to sprinkle lawns during dry weather, steam being used in the place of water. From the top of a closed cauldron rose two vertical pipes, which at their upper ends had short, right-angle bends. Between them was hung a hollow globe, pivoted on two short tubes projecting from its sides into the upright tubes. Two littleL-shaped pipes projected from opposite sides of the globe, at the ends of a diameter, in a plane perpendicular to the axis. On fire being applied to the cauldron, steam was generated. It passed up through the upright, through the pivots, and into the globe, from which it escaped by the twoL-shaped nozzles, causing rapid revolution of the ball. In short, the first steam-engine was a turbine. Curiously enough, we have reverted to this primitive type (scientifically developed, of course) in the most modern engineering practice.

HOW A TURBINE WORKS.

In reciprocating—that is, cylinder—engines steam is admitted into a chamber and the door shut behind it, as it were. As it struggles to expand, it forces out one of the confining walls—that is, the piston—and presently the door opens again, and allows it to escape when it has done its work. In Hero's toy the impact of the issuing molecules against other molecules that have already emerged from the pipes was used. One may compare the reaction to that exerted by a thrown stone on the thrower. If the thrower is standing on skates, the reaction of the stone will cause him to glide backwards, just as if he had pushed off from some fixed object. In the case of the *reaction*—namely, the Hero-type—turbine the nozzle from which the steam or water issues moves, along with bodies to which it may be attached. In *action* turbines steam is led through fixed nozzles or steam-ways, and the momentum of the steam is brought to bear on the surfaces of movable bodies connected with the shaft.

THE DE LAVAL TURBINE.

In its earliest form this turbine was a modification of Hero's. The wheel was merely a pipe bent inS form, attached at its centre to a hollow vertical shaft supplied with steam through a stuffing-box at one extremity. The steam blew out tangentially from the ends of theS, causing the shaft to revolve rapidly and work the machinery (usually a cream separator) mounted on it. This motor proved very suitable for dairy work, but was too wasteful of steam to be useful where high power was needed.

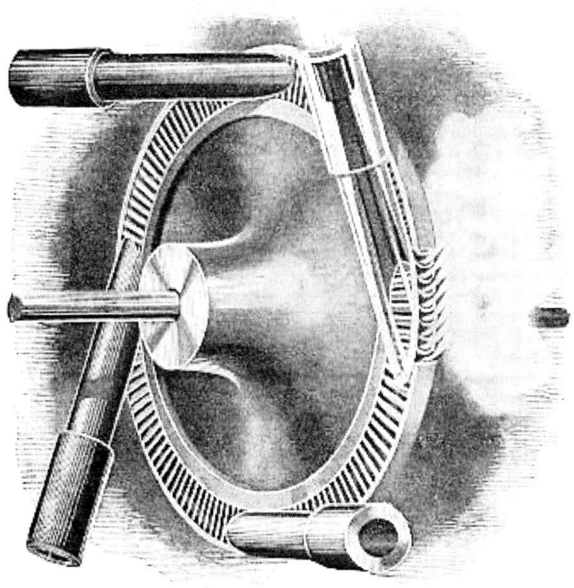

Fig. 36.—The wheel and nozzles of a De Laval turbine.

In the De Laval turbine as now constructed the steam is blown from stationary nozzles against vanes mounted on a revolving wheel. Fig. 36 shows the nozzles and a turbine wheel. The wheel is made as a solid disc, to the circumference of which the vanes are dovetailed separately in a single row. Each vane is of curved section, the concave side directed towards the nozzles, which, as will be gathered from the "transparent" specimen on the right of our illustration, gradually expand towards the mouth. This is to allow the expansion of the steam, and a consequent gain of velocity. As it issues, each molecule strikes against the concave face of a vane, and, while changing its direction, is robbed of its kinetic energy, which passes to the wheel. To turn once more to a stone-throwing comparison, it is as if a boy were pelting the wheel with an enormous number of tiny stones. Now, escaping high-pressure steam moves very fast indeed. To give figures, if it enters the small end of a De Laval nozzle at 200 lbs. per square inch, it will leave the big end at a velocity of 48 miles per *minute*—that is, at a speed which would take it right round the world in 8½ hours! The wheel itself would not move at more than about one-third of this speed as a maximum.[7] But even so, it may make as many as 30,000 revolutions per minute. A mechanical difficulty is now encountered—namely, that arising from vibration. No matter how carefully the turbine wheel may be balanced, it is practically impossible to make its centre of gravity coincide exactly with the central point of the shaft; in other words, the wheel will be a bit—perhaps only a tiny fraction of an ounce—heavier on one side than the other. This want of truth causes vibration, which, at the high speed mentioned, would cause the shaft to knock the bearings in which it revolves to pieces, if—and this is the point—those bearings were close to the wheel M. de Laval mounted the wheel on a shaft long enough between the bearings to "whip," or bend a little, and the difficulty was surmounted.

The normal speed of the turbine wheel is too high for direct driving of some machinery, so it is reduced by means of gearing. To dynamos, pumps, and air-fans it is often coupled direct.

THE PARSONS TURBINE.

At the grand naval review held in 1897 in honour of Queen Victoria's diamond jubilee, one of the most noteworthy sights was the little *Turbinia* of 44½ tons burthen, which darted about among the floating forts at a speed much surpassing that of the fastest "destroyer." Inside the nimble little craft were engines developing 2,000 horse power, without any of the clank and vibration which usually reigns in the engine-room of a high-speed vessel. The *Turbinia* was the first turbine-driven boat, and as such, even apart from her extraordinary pace, she attracted great attention. Since 1897 the Parsons turbine has been installed on many ships, including several men-of-war, and it seems probable that the time is not far distant when reciprocating engines will be abandoned on all high-speed craft.

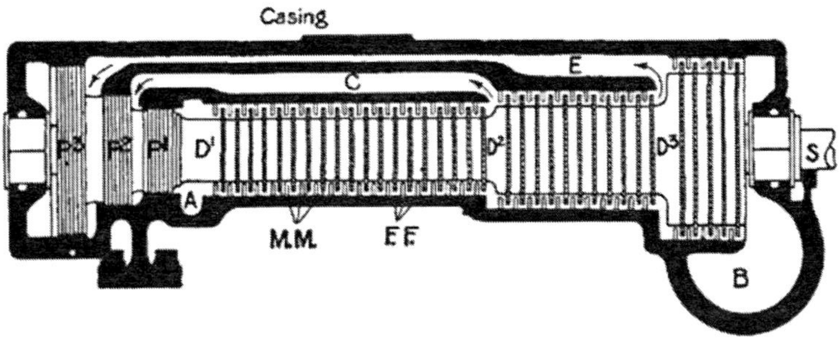

Fig. 37.—Section of a Parsons turbine.

###Insert image036.jpg here###

The essential parts of a Parsons turbine are:—(1) The shaft, on which is mounted (2) the drum; (3) the cylindrical casing inside which the drum revolves; (4) the vanes on the drum and casing; (5) the balance pistons. Fig. 37 shows a diagrammatic turbine in section. The drum, it will be noticed, increases its diameter in three stages,D1,D2,D3, towards the right. From end to end it is studded with little vanes,M M, set in parallel rings small distances apart. Each vane has a curved section (see Fig. 38), the hollow side facing towards the left. The vanes stick out from the drum like short spokes, and their outer ends almost touch the casing. To the latter are attached equally-spaced rings of fixed vanes,F F, pointing inwards towards the drum, and occupying the intervals between the rings of moving vanes. Their concave sides also face towards the left, but, as seen in Fig. 38, their line of curve lies the reverse way to that ofM M. Steam enters the casing atA, and at once rushes through the vanes towards the outlet atB. It meets the first row of fixed vanes, and has its path so deflected that it strikes the ring of moving (or drum) vanes at the most effective angle, and pushes them round. It then has its direction changed by the ring ofF F, so that it may treat the next row ofM Min a similar fashion.

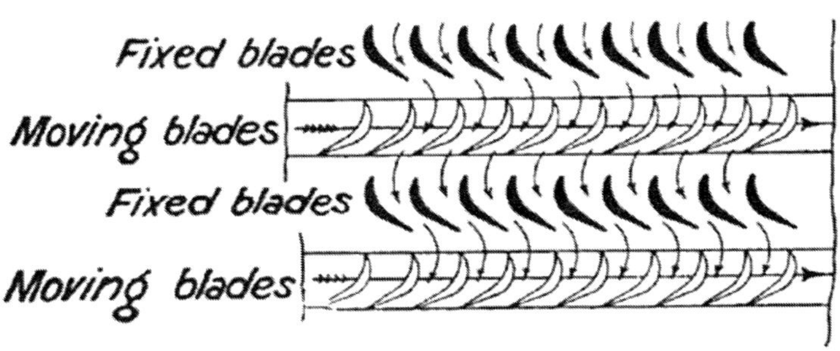

Fig. 38.—Blades or vanes of a Parsons turbine.

One of the low-pressure turbines of the Carmania, in casing. Its size will be inferred from comparison with the man standing near the end of the casing.

THE EXPANSIVE ACTION OF STEAM IN A TURBINE.

On reaching the end ofD1 it enters the second, or intermediate, set of vanes. The drum here is of a greater diameter, and the blades are longer and set somewhat farther apart, to give a freer passage to the now partly expanded steam, which has lost pressure but gained velocity. The process of movement is repeated through this stage; and again inD3, the low-pressure drum. The steam then escapes to the condenser throughB, having by this time expanded very many times; and it is found advisable, for reasons explained in connection with compound steam-engines, to have a separate turbine in an independent casing for the extreme stages of expansion.

The vanes are made of brass. In the turbines of the *Carmania*, the huge Cunard liner, 1,115,000 vanes are used. The largest diameter of the drums is 11 feet, and each low-pressure turbine weighs 350 tons.

BALANCING OF THRUST.

The push exerted by the steam on the blades not only turns the drum, but presses it in the direction in which the steam flows. This end thrust is counterbalanced by means of the "dummy" pistons,P1,P2,P3. Each dummy consists of a number of discs revolving between rings projecting from the casing, the distance between discs and rings being so small that but little steam can pass. In the high-pressure compartment the steam pushesP1 to the left with the same pressure as it pushes the blades ofD1 to the right. After completing the first stage it fills the passageC, which communicates with the second piston,P2, and the pressure on that piston negatives the thrust onD2. Similarly, the passageEcauses the steam to press equally onP3 and the vanes ofD3. So that the bearings in which the shaft revolves have but little thrust to take. This form of compensation is necessary in marine as well as in stationary turbines. In the former the dummy pistons are so proportioned that the forward thrust given by them and the screw combined is almost equal to the thrust aft of the moving vanes.

One of the turbine drums of the Carmania. Note the rows of vanes. The drum is here being tested for perfect balance on two absolutely level supports.

ADVANTAGES OF THE MARINE TURBINE.

(1.) Absence of vibration. Reciprocating engines, however well balanced, cause a shaking of the whole ship which is very unpleasant to passengers. The turbine, on the other hand, being almost perfectly balanced, runs so smoothly at the highest speeds that, if the hand be laid on the covering, it is sometimes almost impossible to tell whether the machinery is in motion. As a consequence of this smooth running there is little noise in the engine-room—a pleasant contrast to the deafening roar of reciprocating engines. (2.) Turbines occupy less room. (3.) They are more easily tended. (4.) They require fewer repairs, since the rubbing surfaces are very small as compared to those of reciprocating engines. (5.) They are more economical at high speeds. It must be remembered that a turbine is essentially meant for high speeds. If run slowly, the steam will escape through the many passages without doing much work.

Owing to its construction, a turbine cannot be reversed like a cylinder engine. It therefore becomes necessary to fit special astern turbines to one or more of the screw shafts, for use when the ship has to be stopped or moved astern. Under ordinary conditions these turbines revolve idly in their cases.

The highest speed ever attained on the sea was the forty-two miles per hour of the unfortunate *Viper*, a turbine destroyer which developed 11,500 horse power, though displacing only 370 tons. This velocity would compare favourably with that of a good many expresses on certain railways that we could name. In the future thirty miles an hour will certainly be attained by turbine-driven liners.

[7] Even at this speed the wheel has a circumferential velocity of two-thirds that of a bullet shot from a Lee-Metford rifle. A vane weighing only 250 grains (about ½ oz.) exerts under these conditions a centrifugal pull of 15 cwt. on the wheel!

CHAPTER IV.
THE INTERNAL-COMBUSTION ENGINE.

The meaning of the term—Action of the internal-combustion engine—The motor car—The starting-handle—The engine—The carburetter—Ignition of the charge—Advancing the spark—Governing the engine—The clutch—The gear-box—The compensating gear—The silencer—The brakes—Speed of cars.

THE MEANING OF THE TERM "INTERNAL-COMBUSTION ENGINE."

IN the case of a steam-boiler the energy of combustion is transmitted to water inside an air-tight vessel. The fuel does not actually touch the "working fluid." In the gas or oil engine the fuel is brought into contact and mixed with the working fluid, which is air. It combines suddenly with it in the cylinder, and heat energy is developed so rapidly that the act is called an explosion. Coal gas, mineral oils, alcohol, petrol, etc., all contain hydrogen and carbon. If air, which contributes oxygen, be added to any of these in due proportion, the mixture becomes highly explosive. On a light being applied, oxygen and carbon unite, also hydrogen and oxygen, and violent heat is generated, causing a violent molecular bombardment of the sides of the vessel containing the mixture. Now, if the mixture be *compressed* it becomes hotter and hotter, until a point is reached at which it ignites spontaneously. Early gas-engines did not compress the charge before ignition. Alphonse Beau de Rochas, a Frenchman, first thought of making the piston of the engine squeeze the mixture before ignition; and from the year 1862, when he proposed this innovation, the success of the internal-combustion engine may be said to date.

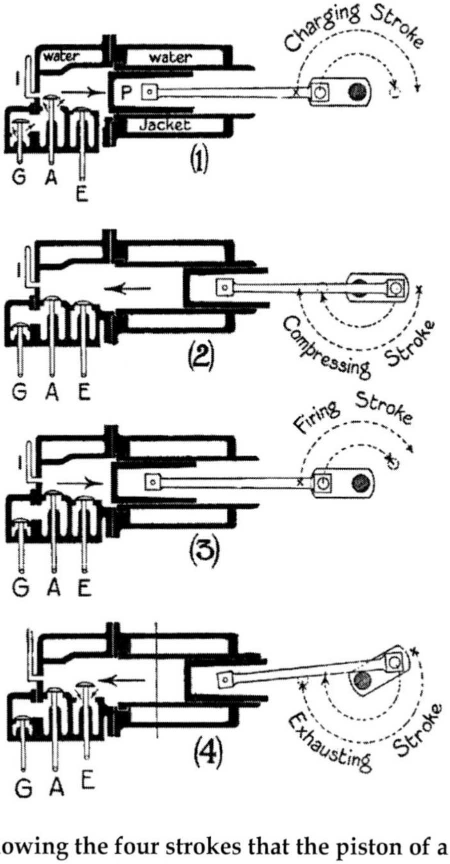

Fig. 39.—Showing the four strokes that the piston of a gas-engine makes during one "cycle."

ACTION OF THE ENGINE.

The gas-engine, the oil-engine, and the motor-car engine are similar in general principles. The cylinder has, instead of a slide-valve, two, or sometimes three, "mushroom" valves, which may be described as small and thick round plates, with bevelled edges, mounted on the ends of short rods, called stems. These valves open into the cylinder, upwards, downwards, or horizontally, as the case may be; being pushed in by cams projecting from a shaft rotated by the engine. For the present we will confine our attention to the series of operations which causes the engine to work. This series is called the Beau de Rochas, or Otto, cycle, and includes four movements of the piston. Reference to Fig. 39 will show exactly what happens in a gas-engine—(1) The piston moves from left to right, and just as the movement commences valvesG(gas) andA(air) open to admit the explosive mixture. By the time thatPhas reached the end of its travel these valves have closed again. (2) The piston returns to the left, compressing the mixture, which has no way of escape open to it. At the end of the stroke the charge is ignited by an incandescent tubeI(in motor car and some stationary engines by an electric spark), and (3) the piston flies out again on the "explosion" stroke. Before it reaches the limit position, valveE(exhaust) opens, and (4) the piston flies back under the momentum of the fly-wheel, driving out the burnt gases through the still openE. The "cycle" is now complete. There has been suction, compression (including ignition), combustion, and exhaustion. It is evident that a heavy fly-wheel must be attached to the crank shaft, because the energy of one stroke (the explosion) has to serve for the whole cycle; in other words, for two complete revolutions of the crank. A single-cylinder steam-engine develops an impulse every half-turn—that is, four times as often. In order to get a more constant turning effect, motor cars have two, three, four, six, and even eight cylinders. Four-cylinder engines are at present the most popular type for powerful cars.

THE MOTOR CAR.

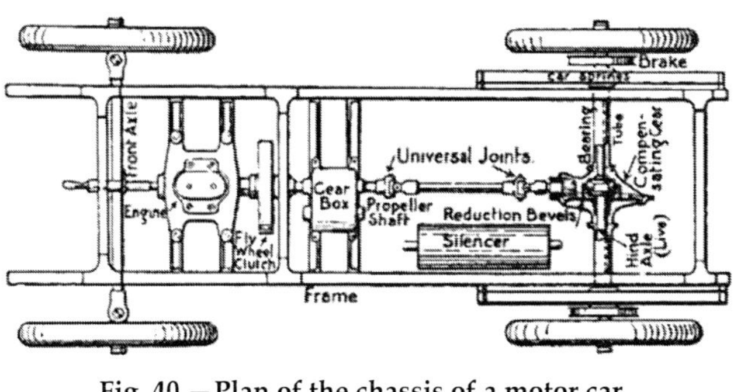

Fig. 40.—Plan of the chassis of a motor car.

We will now proceed to an examination of the motor car, which, in addition to mechanical apparatus for the transmission of motion to the driving-wheels, includes all the fundamental adjuncts of the internal-combustion engine.[8] Fig. 40 is a bird's-eye view of the *chassis* (or "works" and wheels) of a car, from which the body has been removed. Starting at the left, we have the handle for setting the engine in motion; the engine (a two-cylinder in this case); the fly-wheel, inside which is the clutch; the gear-box, containing the cogs for altering the speed of revolution of the driving-wheels relatively to that of the engine; the propeller shaft; the silencer, for deadening the noise of the exhaust; and the bevel-gear, for turning the driving-wheels. In the particular type of car here considered you will notice that a "direct," or shaft, drive is used. The shaft has at each end a flexible, or "universal," joint, which allows the shaft to turn freely, even though it may not be in a line with the shaft projecting from the gear-box. It must be remembered that the engine and gear-box are mounted on the frame, between which and the axles are springs, so that when the car bumps up and down, the shaft describes part of a circle, of which the gear-box end is the centre.

An alternative method of driving is by means of chains, which run round sprocket (cog) wheels on the ends of a shaft crossing the frame just behind the gear-box, and round larger sprockets attached to the hubs of the driving-wheels. In such a case the axles of the driving-wheel are fixed to the springs, and the wheels revolve round them. Where a Cardan (shaft) drive is used the axles are attached rigidly to the wheels at one end, and extend, through tubes fixed to the springs, to bevel-wheels in a central compensating-gear box (of which more presently).

Several parts—the carburetter, tanks, governor, and pump—are not shown in the general plan. These will be referred to in the more detailed account that follows.

THE STARTING-HANDLE.

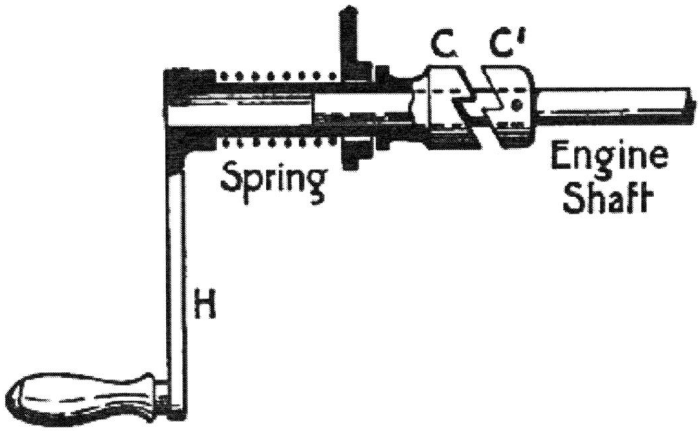

Fig. 41.—The starting-handle.

Fig. 41 gives the starting-handle in part section. The handleHis attached to a tube which terminates in a clutch,C. A powerful spring keepsCnormally apart from a second clutch,C1, keyed to the engine shaft. When the driver wishes to start the engine he presses the handle towards the right, brings the clutches together, and turns the handle in a clockwise direction. As soon as the engine begins to fire, the faces of the clutches slip over one another.

THE ENGINE.

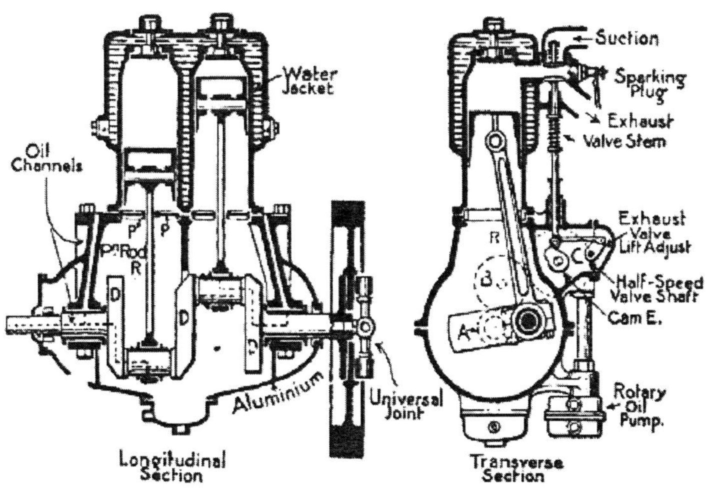

Fig. 42.—End and cross sections of a two-cylinder motor.

We next examine the two-cylinder engine (Fig. 42). Each cylinder is surrounded by a water-jacket, through which water is circulated by a pump[9] (Fig. 43). The heat generated by combustion is so great that the walls of the cylinder would soon become red-hot unless some of the heat were quickly carried away. The pistons are of "trunk" form—that is, long enough to act as guides and absorb the oblique thrust of the piston rods. Three or more piston rings lying in slots (not shown) prevent the escape of gas past the piston. It is interesting to notice that the efficiency of an internal-combustion engine depends so largely on the good fit of these moving parts, that cylinders, pistons, and rings must be exceedingly true. A good firm will turn out standard parts which are well within1/5000 of an inch of perfect truth. It is also a wonderful testimony to the quality of the materials used that, if properly looked after, an engine which has made many millions of revolutions, at the rate of 1,000 to 2,000 per minute, often shows no appreciable signs of wear. In one particular test an engine was run *continuously for several months*, and at the end of the trial was in absolutely perfect condition.

The cranks revolve in an oil-tight case (generally made of aluminium), and dip in oil, which they splash up into the cylinder to keep the piston well lubricated. The plate,P P, through a slot in which the piston rod works, prevents an excess of oil being flung up. Channels are provided for leading oil into the bearings. The cranks are 180° apart. While one piston is being driven out by an explosion, the other is compressing its charge prior to ignition, so that the one action deadens the other. Therefore two explosions occur in one revolution of the cranks, and none during the next revolution. If both cranks were in line, the pistons would move together, giving one explosion each revolution.

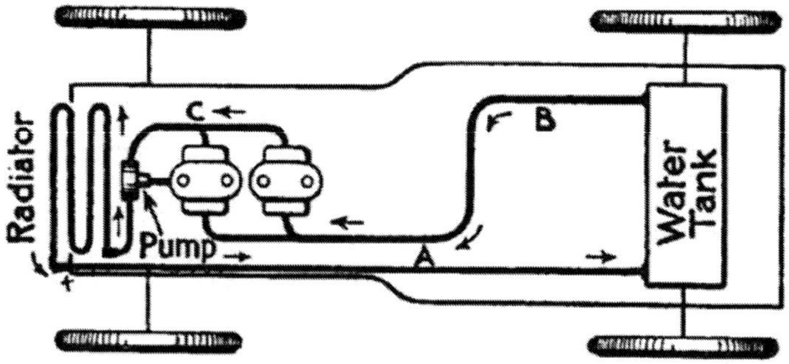

Fig. 43.—Showing how the water which cools the cylinders is circulated.

The valve seats, and the inlet and exhaust pipes, are seen in section. The inlet valve here works automatically, being pulled in by suction; but on many engines—on all powerful engines—the inlet, like the exhaust valve, is lifted by a cam, lest it should stick or work irregularly. Three dotted circles showA, a cog on the crank shaft;B, a "lay" cog, which transmits motion toC, on a short shaft rotating the cam that lifts the exhaust valve.C, having twice as many teeth asA, revolves at half its rate. This ensures that the valve shall be lifted only once in two revolutions of the crank shaft to which it is geared. The cogs are timed, or arranged, so that the cam begins to lift the valve when the piston has made about seven-eighths of its explosion stroke, and closes the valve at the end of the exhaust stroke.

THE CARBURETTER.

A motor car generally uses petrol as its fuel. Petrol is one of the more volatile products of petroleum, and has a specific gravity of about 680—that is, volume for volume, its weight is to that of water in the proportion of 680 to 1,000. It is extremely dangerous, as it gives off an inflammable gas at ordinary temperatures. Benzine, which we use to clean clothes, is practically the same as petrol, and should be treated with equal care. The function of a *carburetter* is to reduce petrol to a very fine spray and mix it with a due quantity of air. The device consists of two main parts (Fig. 44)—the *float chamber* and the *jet chamber*. In the former is a contrivance for regulating the petrol supply. A float—a cork, or air-tight metal box—is arranged to move freely up and down the stem of a needle-valve, which closes the inlet from the tank. At the bottom of the chamber are two pivoted levers,W W, which, when the float rests on them, tip up and lift the valve. Petrol flows in and raises the float. This allows the valve to sink and cut off the supply. If the valve is a good fit and the float is of the correct weight, the petrol will never rise higher than the tip of the jetG.

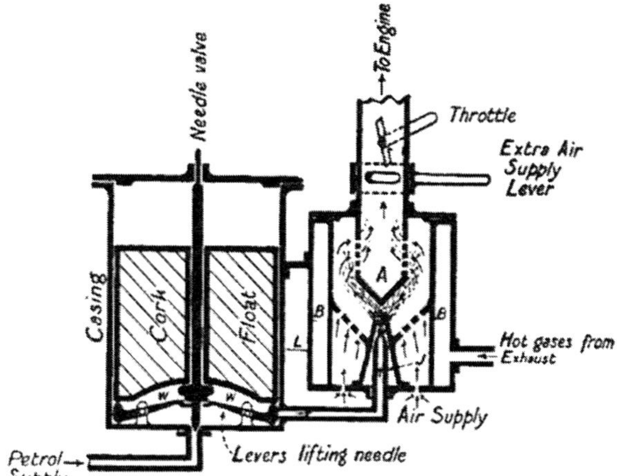

Fig. 44.—Section of a carburetter.

The suction of the engine makes petrol spirt through the jet (which has a very small hole in its end) and atomize itself against a spraying-cone,A. It then passes to the engine inlet pipe through a number of openings, after mixing with air entering from below. An extra air inlet, controllable by the driver, is generally added, unless the carburetter be of a type which automatically maintains constant proportions of air and vapour. The jet chamber is often surrounded by a jacket, through which part of the hot exhaust gases circulate. In cold weather especially this is a valuable aid to vaporization.

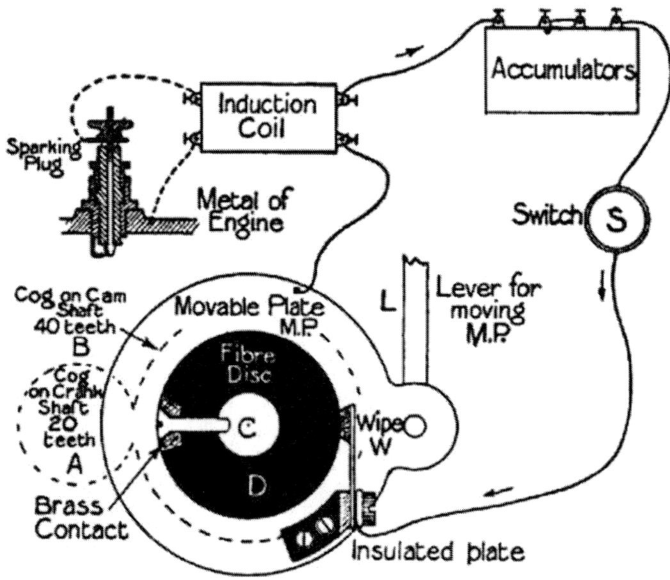

Fig. 45.—Sketch of the electrical ignition arrangements on a motor car.

IGNITION OF THE CHARGE.

All petrol-cars now use electrical ignition. There are two main systems—(1) by an accumulator and induction coil; (2) *magneto ignition,* by means of a small dynamo driven by the engine. A general arrangement of the first is shown in Fig. 45. A disc,D, of some insulating material—fibre or vulcanite—is mounted on the cam, or half-speed, shaft. Into the circumference is let a piece of brass, called the contact-piece, through which a screw passes to the cam shaft. A movable plate,M P, which can be rotated concentrically withDthrough part of a circle, carries a "wipe" block at the end of a spring, which presses it againstD. The spring itself is attached to an insulated plate. When the revolution ofDbrings the wipe and contact together, current flows from the accumulator through switchSto the wipe; through the contact-piece toC; fromCtoM Pand the induction coil; and back to the accumulator. This is the *primary, or low-tension, circuit*. A *high-tension* current is induced by the coil in the *secondary* circuit, indicated by dotted lines.[10] In this circuit is the sparking-plug (see Fig. 46), having a central insulated rod in connection with one terminal of the secondary coil. Between it and a bent wire projecting from the iron casing of the plug (in contact with the other terminal of the secondary coil through the metal of the engine, to which one wire of the circuit is attached) is a small gap, across which the secondary current leaps when the primary current is broken by the wipe and contact parting company. The spark is intensely hot, and suffices to ignite the compressed charge in the cylinder.

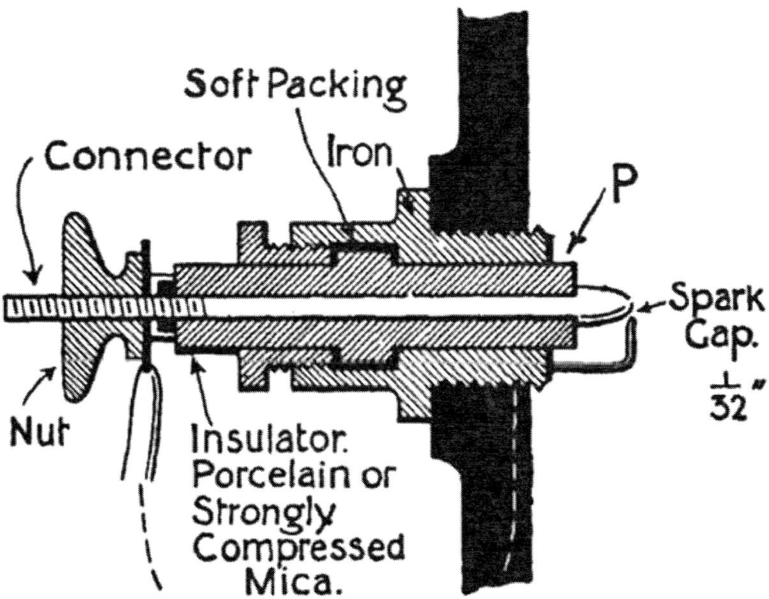

Fig. 46.—Section of a sparking-plug.

ADVANCING THE SPARK.

We will assume that the position ofW(in Fig. 45) is such that the contact touchesWat the moment when the piston has just completed the compression stroke. Now, the actual combustion of the charge occupies an appreciable time, and with the engine running at high speed the piston would have travelled some way down the cylinder before the full force of the explosion was developed. But by raising leverL, the position ofWmay be so altered that contact is made slightly *before* the compression stroke is complete, so that the charge is fairly alight by the time the piston has altered its direction. This is called *advancing* the spark.

GOVERNING THE ENGINE.

There are several methods of controlling the speed of internal-combustion engines. The operating mechanism in most cases is a centrifugal ball-governor. When the speed has reached the fixed limit it either (1) raises the exhaust valve, so that no fresh charges are drawn in; (2) prevents the opening of the inlet valve; or (3) throttles the gas supply. The last is now most commonly used on motor cars, in conjunction with some device for putting it out of action when the driver wishes to exceed the highest speed that it normally permits.

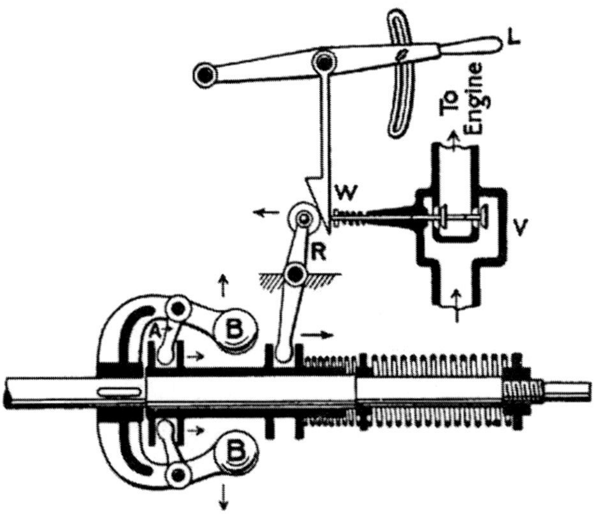

Fig. 47.—One form of governor used on motor cars.

A sketch of a neat governor, with regulating attachment, is given in Fig. 47. The governor shaft is driven from the engine. As the balls,B B, increase their velocity, they fly away from the shaft and move the arms,A A, and a sliding tube,C, towards the right. This rocks the leverR, and allows the valves in the inlet pipe to close and reduce the supply of air and gas. A wedge,W, which can be raised or lowered by leverL, intervenes between the end ofRand the valve stem. If this lever be lifted to its highest position, the governing commences at a lower speed, as the valve then has but a short distance to travel before closing completely. For high speeds the driver depressesL, forces the wedge down, and so minimizes the effect of the governor.

THE CLUTCH.

The engine shaft has on its rear end the fly-wheel, which has a broad and heavy rim, turned to a conical shape inside. Close to this, revolving loosely on the shaft, is the clutch plate, a heavy disc with a broad edge so shaped as to fit the inside of a fly-wheel. It is generally faced with leather. A very strong spring presses the plate into the fly-wheel, and the resulting friction is sufficient to prevent any slip. Projections on the rear of the clutch engage with the gear-box shaft. The driver throws out the clutch by depressing a lever with his foot. Some clutches dispense with the leather lining. These are termed *metal to metal* clutches.

THE GEAR-BOX.

We now come to a very interesting detail of the motor car, the gear-box. The steam-engine has its speed increased by admitting more steam to the cylinders. But an explosion engine must be run at a high speed to develop its full power, and when heavier work has to be done on a hill it becomes necessary to alter the speed ratio of engine to driving-wheels. Our illustration (Fig. 48) gives a section of a gear-box, which will serve as a typical example. It provides three forward speeds and one reverse. To understand how it works, we must study the illustration carefully. Pinion 1 is mounted on a hollow shaft turned by the clutch. Into the hollow shaft projects the end of another shaft carrying pinions 6 and 4. Pinion 6 slides up and down this shaft, which is square at this point, but round inside the *loose* pinion 4. Pinions 2 and 3 are keyed to a square secondary shaft, and are respectively always in gear with 1 and 4; but 5 can be slid backwards and forwards so as to engage or disengage with 6. In the illustration no gear is "in." If the engine is working, 1 revolves 2, 2 turns 3, and 3 revolves 4 idly on its shaft.

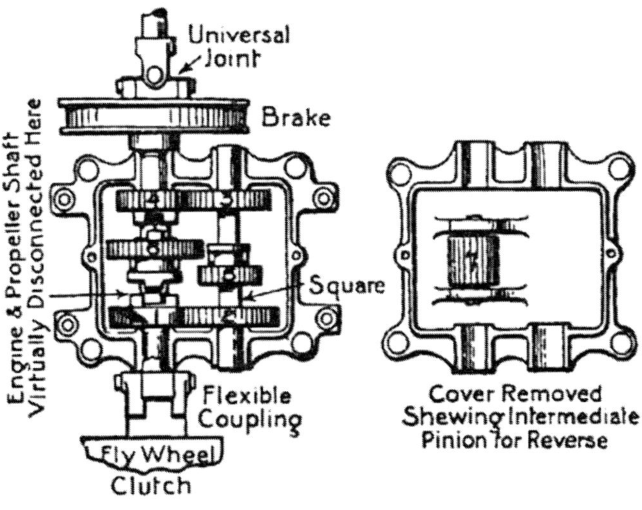

Fig. 48.—The gear-box of a motor car.

To get the lowest, or "first," speed the driver moves his lever and slides 5 into gear with 6. The transmission then is: 1 turns 2, 2 turns 5, 5 turns 6, 6 turns the propeller shaft through the universal joint. For the second speed, 5 and 6 are disengaged, and 6 is moved up the page, as it were, till projections on it interlock with slots in 4; thus driving 1, 2, 3, 4, shaft. For the third, or "solid," speed, 6 is pulled down into connection with 1, and couples the engine shaft direct to the propeller shaft.

The "reverse" is accomplished by raising a long pinion, 7, which lies in the gear-box under 5 and 6. The drive then is 1, 2, 5, 7, 6. There being an odd number of pinions now engaged, the propeller shaft turns in the reverse direction to that of the engine shaft.

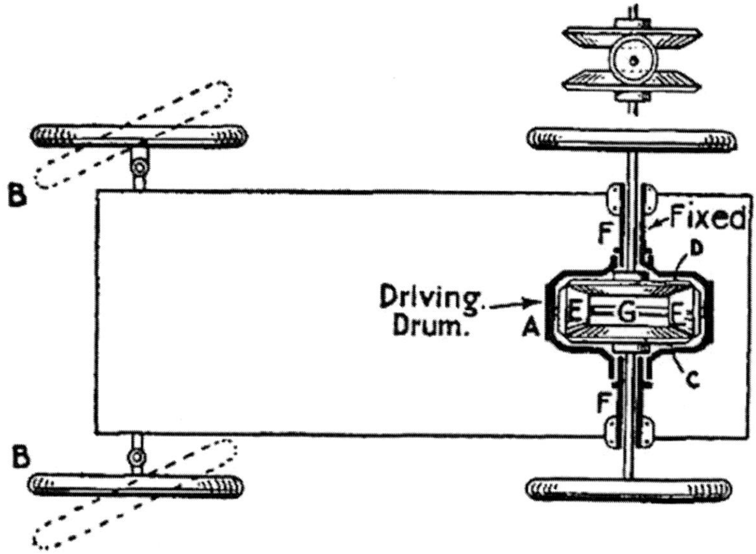

Fig. 49.

THE COMPENSATING GEAR.

Every axle of a railway train carries a wheel at each end, rigidly attached to it. When rounding a corner the outside wheel has further to travel than the other, and consequently one or both wheels must slip. The curves are made so gentle, however, that the amount of slip is very small. But with a traction-engine, motor car, or tricycle the case is different, for all have to describe circles of very small diameter in proportion to the length of the vehicle. Therefore in every case a *compensating gear* is fitted, to allow the wheels to turn at different speeds, while permitting them both to drive. Fig. 49 is an exaggerated sketch of the gear. The axles of the moving wheels turn inside tubes attached to the springs and a central casing (not shown), and terminate in large bevel-wheels,CandD. Between these are small bevels mounted on a shaft supported by the driving drum. If the latter be rotated, the bevels would turnCandDat equal speeds, assuming that both axles revolve without friction in their bearings. We will suppose that the drum is turned 50 times a minute. Now, if one wheel be held, the other will revolve 100 times a minute; or, if one be slowed, the other will increase its speed by a corresponding amount. The *average* speed remains 50. It should be mentioned that drumAhas incorporated with it on the outside a bevel-wheel (not shown) rotated by a smaller bevel on the end of the propeller shaft.

THE SILENCER.

The petrol-engine, as now used, emits the products of combustion at a high pressure. If unchecked, they expand violently, and cause a partial vacuum in the exhaust pipe, into which the air rushes back with such violence as to cause a loud noise. Devices called *silencers* are therefore fitted, to render the escape more gradual, and split it up among a number of small apertures. The simplest form of silencer is a cylindrical box, with a number of finely perforated tubes passing from end to end of it. The exhaust gases pouring into the box maintain a constant pressure somewhat higher than that of the atmosphere, but as the gases are escaping from it in a fairly steady stream the noise becomes a gentle hiss rather than a "pop." There are numerous types of silencers, but all employ this principle in one form or another.

THE BRAKES.

Every car carries at least two brakes of band pattern—one, usually worked by a side hand-lever, acting on the axle or hubs of the driving-wheel; the other, operated by the foot, acting on the transmission gear (see Fig. 48). The latter brake is generally arranged to withdraw the clutch simultaneously. Tests have proved that even heavy cars can be pulled up in astonishingly short distances, considering their rate of travel. Trials made in the United States with a touring car and a four-in-hand coach gave 25⅓ and 70 feet respectively for the distance in which the speed could be reduced from sixteen miles per hour to zero.

SPEED OF CARS.

As regards speed, motor cars can rival the fastest express trains, even on long journeys. In fact, feats performed during the Gordon-Bennett and other races have equalled railway performances over equal distances. When we come to record speeds, we find a car, specially built for the purpose, covering a mile in less than half a minute. A speed of over 120 miles an hour has actually been reached. Engines of 150 h.p. can now be packed into a vehicle scaling less than 1½ tons. Even on touring cars are often found engines developing

40 to 60 h.p., which force the car up steep hills at a pace nothing less than astonishing. In the future the motor car will revolutionize our modes of life to an extent comparable to the changes effected by the advent of the steam-engine. Even since 1896, when the "man-with-the-flag" law was abolished in the British Isles, the motor has reduced distances, opened up country districts, and generally quickened the pulses of the community in a manner which makes it hazardous to prophesy how the next generation will live.

Note.—The author is much indebted to Mr. Wilfrid J. Lineham, M. Inst. C.E., for several of the illustrations which appear in the above chapter.

[8] Steam-driven cars are not considered in this chapter, as their principle is much the same as that of the ordinary locomotive.

[9] On some cars natural circulation is used, the hot water flowing from the top of the cylinder to the tank, from which it returns, after being cooled, to the bottom of the cylinder.

[10] For explanation of the induction coil.

CHAPTER V.
ELECTRICAL APPARATUS.

What is electricity?—Forms of electricity—Magnetism—The permanent magnet—Lines of force—Electro-magnets—The electric bell—The induction coil—The condenser—Transformation of current—Uses of the induction coil.

WHAT IS ELECTRICITY?

OF the ultimate nature of electricity, as of that of heat and light, we are at present ignorant. But it has been clearly established that all three phenomena are but manifestations of the energy pervading the universe. By means of suitable apparatus one form can be converted into another form. The heat of fuel burnt in a boiler furnace develops mechanical energy in the engine which the boiler feeds with steam. The engine revolves a dynamo, and the electric current thereby generated can be passed through wires to produce mechanical motion, heat, or light. We must remain content, therefore, with assuming that electricity is energy or motion transmitted through the ether from molecule to molecule, or from atom to atom, of matter. Scientific investigation has taught us how to produce it at will, how to harness it to our uses, and how to measure it; but not *what* it is. That question may, perhaps, remain unanswered till the end of human history. A great difficulty attending the explanation of electrical action is this—that, except in one or two cases, no comparison can be established between it and the operation of gases and fluids. When dealing with the steam-engine, any ordinary intelligence soon grasps the principles which govern the use of steam in cylinders or turbines. The diagrams show, it is hoped, quite plainly "how it works." But electricity is elusive, invisible; and the greatest authorities cannot say what goes on at the poles of a magnet or on the surface of an electrified body. Even the existence of "negative" and "positive" electricity is problematical. However, we see the effects, and we know that if one thing is done another thing happens; so that we are at least able to use terms which, while convenient, are not at present controverted by scientific progress.

FORMS OF ELECTRICITY.

Rub a vulcanite rod and hold one end near some tiny pieces of paper. They fly to it, stick to it for a time, and then fall off. The rod was electrified—that is, its surface was affected in such a way as to be in a state of molecular strain which the contact of the paper fragments alleviated. By rubbing large surfaces and collecting the electricity in suitable receivers the strain can be made to relieve itself in the form of a violent discharge accompanied by a bright flash. This form of electricity is known as *static*.

Next, place a copper plate and a zinc plate into a jar full of diluted sulphuric acid. If a wire be attached to them a current of electricity is said to *flow* along the wire. We must not, however, imagine that anything actually moves along inside the wire, as water, steam, or air, passes through a pipe. Professor Trowbridge says,[11] "No other agency for transmitting power can be stopped by such slight obstacles as electricity. A thin sheet of paper placed across a tube conveying compressed air would be instantly ruptured. It would take a wall of steel at least an inch thick to stand the pressure of steam which is driving a 10,000 horse-power engine. A thin layer of dirt beneath the wheels of an electric car can prevent the current which propels the car from passing to the rail, and then back to the power-house." There would, indeed, be a puncture of the paper if the current had a sufficient voltage, or pressure; yet the fact remains that *current* electricity can be very easily confined to its conductor by means of some insulating or nonconducting envelope.

MAGNETISM.

The most familiar form of electricity is that known as magnetism. When a bar of steel or iron is magnetized, it is supposed that the molecules in it turn and arrange themselves with all their north-seeking poles towards the one end of the bar, and their south-seeking poles towards the other. If the bar is balanced freely on a pivot, it comes to rest pointing north and south; for, the earth being a huge magnet, its north pole attracts all the north-seeking poles of the molecules, and its south poles the south-seeking poles. (The north-*seeking* pole of a magnet is marked N., though it is in reality the *south* pole; for unlike poles are mutually attractive, and like poles repellent.)

There are two forms of magnet— *permanent* and *temporary*. If steel is magnetized, it remains so; but soft iron loses practically all its magnetism as soon as the cause of magnetization is withdrawn. This is what we should expect; for steel is more closely compacted than iron, and the molecules therefore would be able to turn about more easily.[12] It is fortunate for us that this is so, since on the rapid magnetization and demagnetization of soft iron depends the action of many of our electrical mechanisms.

THE PERMANENT MAGNET.

Magnets are either (1) straight, in which case they are called bar magnets; or (2) of horseshoe form, as in Figs. 50 and 51. By bending the magnet the two poles are brought close together, and the attraction of both may be exercised simultaneously on a bar of steel or iron.

LINES OF FORCE.

In Fig. 50 are seen a number of dotted lines. These are called *lines of magnetic force*. If you lay a sheet of paper on a horseshoe magnet and sprinkle it with iron dust, you will at once notice how the particles arrange themselves in curves similar in shape to those shown in the illustration. It is supposed (it cannot be *proved*) that magnetic force streams away from the N. pole and describes a circular course through the air back to the S. pole. The same remark applies to the bar magnet.

ELECTRICAL MAGNETS

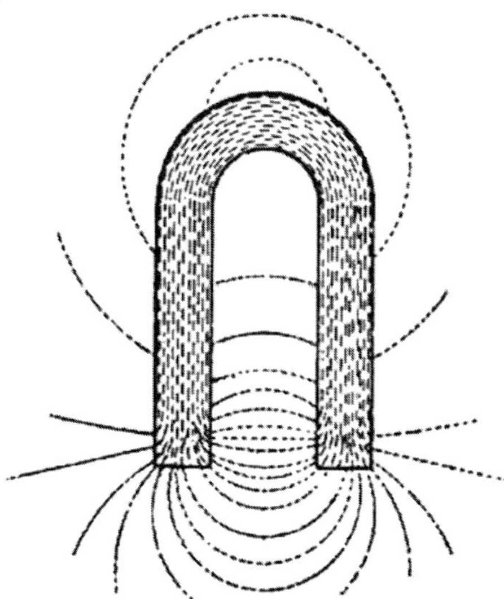

Fig. 50.—Permanent magnet, and the "lines of force" emanating from it.

If an insulated wire is wound round and round a steel or iron bar from end to end, and has its ends connected to the terminals of an electric battery, current rotates round the bar, and the bar is magnetized. By increasing the strength and volume of the current, and multiplying the number of turns of wire, the attractive force of the magnet is increased. Now disconnect the wires from the battery. If of iron, the magnet at once loses its attractive force; but if of steel, it retains it in part. Instead of a simple horseshoe-shaped bar, two shorter bars riveted into a plate are generally used for electromagnets of this type. Coils of wire are wound round each bar, and connected so as to form one continuous whole; but the wire of one coil is wound in the direction opposite to that of the other. The free end of each goes to a battery terminal.

In Fig. 51 you will notice that some of the "lines of force" are deflected through the iron barA. They pass more easily through iron than through air; and will choose iron by preference. The attraction exercised by a magnet on iron may be due to the effort of the lines of force to shorten their paths. It is evident that the closerAcomes to the poles of the magnet the less will be the distance to be travelled from one pole to the bar, along it, and back to the other pole.

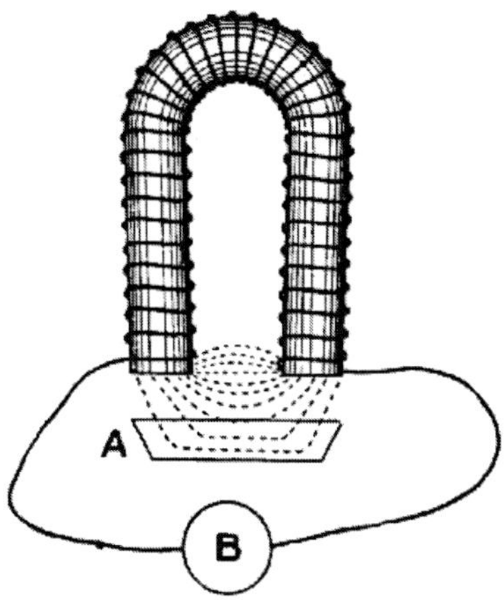

Fig. 51.—Electro-magnet:A, armature;B, battery.

Having now considered electricity in three of its forms—static, current, and rotatory—we will pass to some of its applications.

THE ELECTRIC BELL.

A fit device to begin with is the Electric Bell, which has so largely replaced wire-pulled bells. These last cause a great deal of trouble sometimes, since if a wire snaps it may be necessary to take up carpets and floor-boards to put things right. Their installation is not simple, for at every corner must be put a crank to alter the direction of the pull, and the cranks mean increased friction. But when electric wires have once been properly installed, there should be no need for touching them for an indefinite period. They can be taken round as many corners as you wish without losing any of their conductivity, and be placed wherever is most convenient for examination. One bell may serve a large number of rooms if an *indicator* be used to show where the call was made from, by a card appearing in one of a number of small windows. Before answering a call, the attendant presses in a button to return the card to its normal position.

In Fig. 52 we have a diagrammatic view of an electric bell and current. When the bell-push is pressed in, current flows from the battery to terminalT1, round the electro-magnetM, through the pillarPand flat steel springsSandB, through the platinum-pointed screw, and back to the battery through the push. The circulation of current magnetizesM, which attracts the iron armatureAattached to the springS, and draws the hammerHtowards the gong. Just before the stroke occurs, the springBleaves the tip of the screw, and the circuit is broken, so that the magnet no longer attracts.His carried by its momentum against the gong, and is withdrawn by the spring, untilBonce more makes contact, and the magnet is re-excited. The hammer vibrations recur many times a second as long as the push is pressed in.

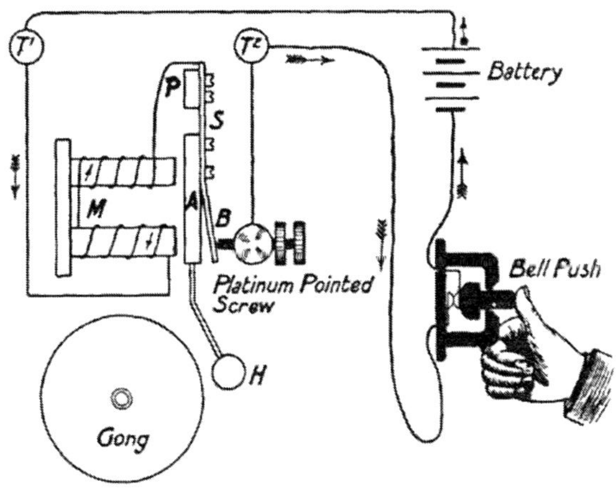

Fig. 52.—Sketch of an electric-bell circuit.

The electric bell is used for so many purposes that they cannot all be noted. It plays an especially important part in telephonic installations to draw the attention of the subscribers, forms an item in automatic fire and burglar alarms, and is a necessary adjunct of railway signalling cabins.

THE INDUCTION OR RUHMKORFF COIL.

Reference was made in connection with the electrical ignition of internal-combustion engines to the *induction coil*. This is a device for increasing the *voltage*, or pressure, of a current. The two-cell accumulator carried in a motor car gives a voltage (otherwise called electro-motive force = E.M.F.) of 4·4 volts. If you attach a wire to one terminal of the accumulator and brush the loose end rapidly across the other terminal, you will notice that a bright spark passes between the wire and the terminal. In reality there are two sparks, one when they touch, and another when they separate, but they occur so closely together that the eye cannot separate the two impressions. A spark of this kind would not be sufficiently hot to ignite a charge in a motor cylinder, and a spark from the induction coil is therefore used.

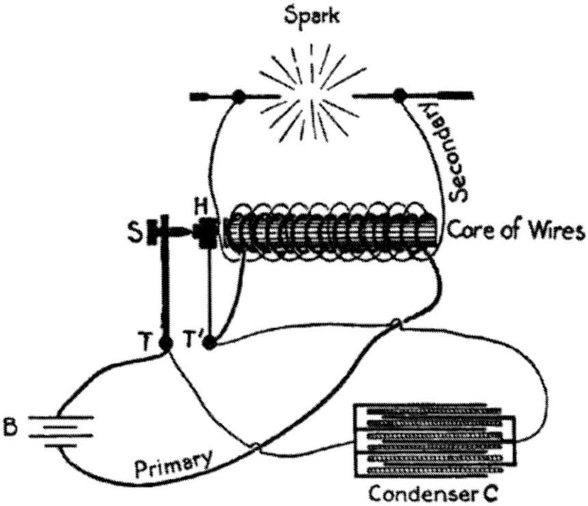

Fig. 53.—Sketch of an induction coil.

We give a sketch of the induction coil in Fig. 53. It consists of a core of soft iron wires round which is wound a layer of coarse insulated wire, denoted by the thick line. One end of the winding of this *primary* coil is attached to the battery, the other to the base of a hammer,H, vibrating between the end of the core and a screw,S, passing through an upright,T, connected with the other terminal of the battery. The action of the hammer is precisely the same as that of the armature of an electric bell. Outside the primary coil are wound many turns of a much finer wire completely insulated from the primary coil. The ends of this *secondary* coil are attached to the objects (in the case of a motor car, the insulated wire of the sparking-plug and a wire projecting from its outer iron casing) between which a spark has to pass. As soon asHtouchesSthe circuit is completed. The core becomes a powerful magnet with external lines of force passing from one pole to the other over and among the turns of the secondary coil.His almost instantaneously attracted by the core, and the break occurs. The lines of force now (at least so it is supposed) sink into the core, cutting through the turns of the "secondary," and causing a powerful current to flow through them. The greater the number of turns, the greater the number of times the lines of force are cut, and the stronger is the current. If sufficiently intense, it jumps any gap in the secondary circuit, heating the intermediate air to a state of incandescence.

THE CONDENSER.

The sudden parting ofHandSwould produce strong sparking across the gap between them if it were not for the condenser, which consists of a number of tinfoil sheets separated by layers of paraffined paper. All the "odd" sheets are connected withT, all the "even" withT1. Now, the more rapid the extinction of magnetism in the core after "break" of the primary circuit, the more rapidly will the lines of force collapse, and the more intense will be the induced current in the secondary coil. The condenser diminishes the period of extinction very greatly, while lengthening the period of magnetization after the "make" of the primary current, and so decreasing the strength of the reverse current.

TRANSFORMATION OF CURRENT.

The difference in the voltage of the primary and secondary currents depends on the length of the windings. If there are 100 turns of wire in the primary, and 100,000 turns in the secondary, the voltage will be increased 1,000 times; so that a 4-volt current is "stepped up" to 4,000 volts. In the largest induction coils the secondary winding absorbs 200–300 miles of wire, and the spark given may be anything up to four feet in length. Such a spark would pierce a glass plate two inches thick.

It must not be supposed that an induction coil increases the *amount* of current given off by a battery. It merely increases its pressure at the expense of its volume—stores up its energy, as it were, until there is enough to do what a low-tension flow could not effect. A fair comparison would be to picture the energy of the low-tension current as the momentum of a number of small pebbles thrown in succession at a door, say 100 a minute. If you went on pelting the door for hours you might make no impression on it, but if you could knead every 100 pebbles into a single stone, and throw these stones one per minute, you would soon break the door in.

Any intermittent current can be transformed as regards its intensity. You may either increase its pressure while decreasing its rate of flow, or *amperage*; or decrease its pressure and increase its flow. In the case that we have considered, a continuous battery current is rendered intermittent by a mechanical contrivance. But if the current comes from an "alternating" dynamo—that is, is already intermittent—the contact-breaker is not needed. There will be more to say about transformation of current in later paragraphs.

USES OF THE INDUCTION COIL.

The induction coil is used—(1.) For passing currents through glass tubes almost exhausted of air or containing highly rarefied gases. The luminous effects of these "Geissler" tubes are very beautiful. (2.) For producing the now famous X or Röntgen rays. These rays accompany the light rays given off at the negative terminal (cathode) of a vacuum tube, and are invisible to the eye unless caught on a fluorescent screen, which reduces their rate of vibration sufficiently for the eye to be sensitive to them. The Röntgen rays have the peculiar property of penetrating many substances quite opaque to light, such as metals, stone, wood, etc., and as a consequence have proved of great use to the surgeon in localizing or determining the nature of an internal injury. They also have a deterrent effect upon cancerous growths. (3.) In wireless telegraphy, to cause powerful electric oscillations in the ether. (4.) On motor cars, for igniting the cylinder charges. (5.) For electrical massage of the body.

[11] "What is Electricity?".

[12] If a magnetized bar be heated to white heat and tapped with a hammer it loses its magnetism, because the distance between the molecules has increased, and the molecules can easily return to their original positions.

CHAPTER VI.
THE ELECTRIC TELEGRAPH.

Needle instruments—Influence of current on the magnetic needle—Method of reversing the current—Sounding instruments—Telegraphic relays—Recording telegraphs—High-speed telegraphy.

TAKE a small pocket compass and wind several turns of fine insulated wire round the case, over the top and under the bottom. Now lay the compass on a table, and turn it about until the coil is on a line with the needle—in fact, covers it. Next touch the terminals of a battery with the ends of the wire. The needle at once shifts either to right or left, and remains in that position as long as the current flows. If you change the wires over, so reversing the direction of the current, the needle at once points in the other direction. It is to this conduct on the part of a magnetic needle when in a "magnetic field" that we owe the existence of the needle telegraph instrument.

NEEDLE INSTRUMENTS.

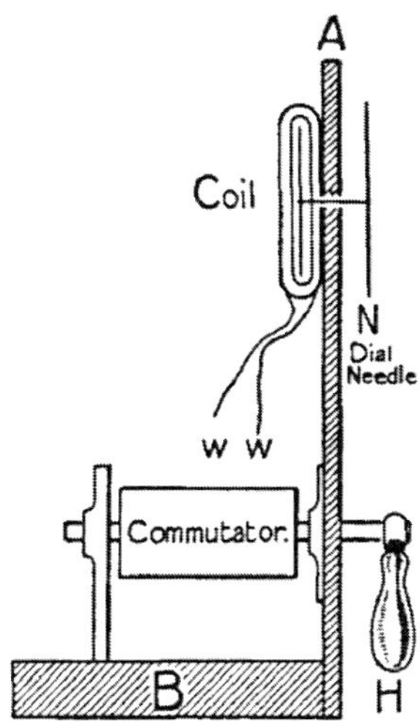

Fig. 54.—Sketch of the side elevation of a Wheatstone needle instrument.

Probably the best-known needle instrument is the Cooke-Wheatstone, largely used in signal-boxes and in some post-offices. A vertical section of it is shown in Fig. 54. It consists of a base,B, and an upright front,A, to the back of which are attached two hollow coils on either side of a magnetic needle mounted on the same shaft as a second dial needle,N, outside the front. The wiresW Ware connected to the telegraph line and to the commutator, a device which, when the operator moves the handleHto right and left, keeps reversing the direction of the current. The needles on both receiving and transmitting instruments wag in accordance with the movements of the handle. One or more movements form an alphabetical letter of the Morse code. Thus, if the needle points first to left, and then to right, and comes to rest in a normal position for a moment, the letter A is signified; right-left-left-left in quick succession = B; right-left-right-left = C, and so on. Where a marking instrument is used, a dot signifies a "left," and a dash a right; and if a "sounder" is employed, the operator judges by the length of the intervals between the clicks.

INFLUENCE OF CURRENT ON A MAGNETIC NEEDLE.

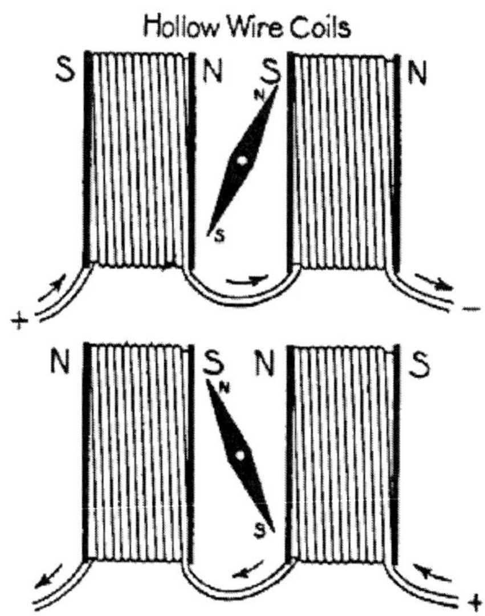

Figs. 55, 56.—The coils of a needle instrument.
The arrows show the direction taken by the current.

Figs. 55 and 56 are two views of the coils and magnetic needle of the Wheatstone instrument as they appear from behind. In Fig. 55 the current enters the left-hand coil from the left, and travels round and round it in a clockwise direction to the other end, whence it passes to the other coil and away to the battery. Now, a coil through which a current passes becomes a magnet. Its polarity depends on the direction in which the current flows. Suppose that you are looking through the coil, and that the current enters it from your end. If the wire is wound in a clockwise direction, the S. pole will be nearest you; if in an anti-clockwise direction, the N. pole. In Fig. 55 the N. poles are at the right end of the coils, the S. poles at the left end; so the N. pole of the needle is attracted to the right, and the S. pole to the left. When the current is reversed, as in Fig. 56, the needle moves over. If no current passes, it remains vertical.

METHOD OF REVERSING THE CURRENT.

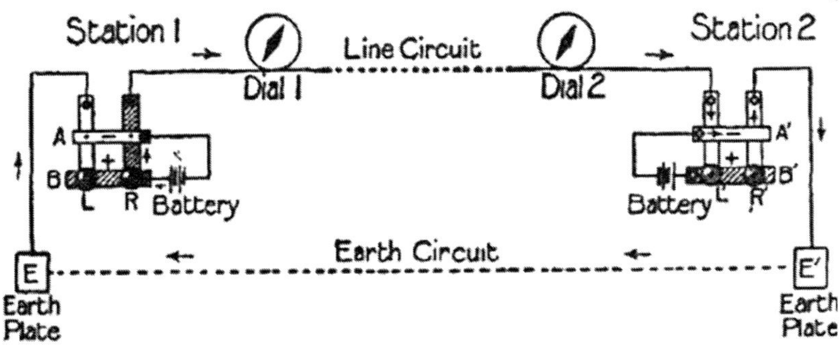

Fig. 57.—General arrangement of needle-instrument circuit. The shaded plates on the left (BandR) are in contact.

A simple method of changing the direction of the current in a two-instrument circuit is shown diagrammatically in Fig. 57. The *principle* is used in the Wheatstone needle instrument. The battery terminals at each station are attached to two brass plates,A B,A1 B1. Crossing these at right angles (underA A1 and overB B1) are the flat brass springs,L R,L1 R1, having buttons at their lower ends, and fixed at their upper ends to baseboards. When at rest they all press upwards against the platesAandA1 respectively.RandL1 are connected with the line circuit, in which are the coils of dials 1 and 2, one at each station.LandR1 are connected with the earth-platesE E1. An operator at station 1 depressesRso as to touchB. Current now flows from the battery toB, thence throughRto the line circuit, round the coils of both dials throughL1 A1 andRto earth-plateE1, through the earth toE, and then back to the battery throughLandA. The needles assume the position shown. To reverse the current the operator allowsRto rise into contact withA, and depressesLto touchB. The course can be traced out easily.

In the Wheatstone "drop-handle" instrument (Fig. 54) the commutator may be described as an insulated core on which are two short lengths of brass tubing. One of these has rubbing against it a spring connected with the terminal of the battery; the other has similar communication with the – terminal. Projecting from each tube is a spike, and rising from the baseboard are four upright brass strips not quite touching the commutator. Those on one side lead to the line circuit, those on the other to the earth-plate. When the handle is turned one way, the spikes touch the forward line strip and the rear earth strip, and *vice versâ* when moved in the opposite direction.

SOUNDING INSTRUMENTS.

Sometimes little brass strips are attached to the dial plate of a needle instrument for the needle to strike against. As these give different notes, the operator can comprehend the message by ear alone. But the most widely used sounding instrument is the Morse sounder, named after its inventor. For this a reversible current is not needed. The receiver is merely an electro-magnet (connected with the line circuit and an earth-plate) which, when a current passes, attracts a little iron bar attached to the middle of a pivoted lever. The free end of the lever works between two stops. Every time the circuit is closed by the transmitting key at the sending station the lever flies down against the lower stop, to rise again when the circuit is broken. The duration of its stay decides whether a "long" or "short" is meant.

TELEGRAPHIC RELAYS.

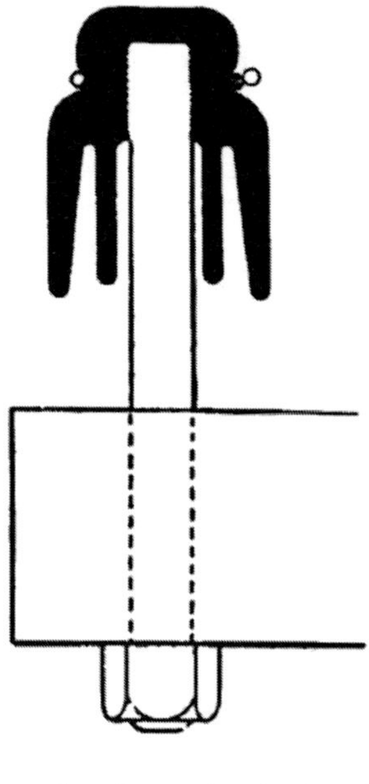

Fig. 58.—Section of a telegraph wire insulator on its arm.
The shaded circle is the line wire, the two blank circles indicate the wire which

ties the line wire to the insulator.

When an electric current has travelled for a long distance through a wire its strength is much reduced on account of the resistance of the wire, and may be insufficient to cause the electro-magnet of the sounder to move the heavy lever. Instead, therefore, of the current acting directly on the sounder magnet, it is used to energize a small magnet, or *relay*, which pulls down a light bar and closes a second "local" circuit—that is, one at the receiver end—worked by a separate battery, which has sufficient power to operate the sounder.

RECORDING TELEGRAPHS.

By attaching a small wheel to the end of a Morse-sounder lever, by arranging an ink-well for the wheel to dip into when the end falls, and by moving a paper ribbon slowly along for the wheel to press against when it rises, a self-recording Morse inker is produced. The ribbon-feeding apparatus is set in motion automatically by the current, and continues to pull the ribbon along until the message is completed.

The Hughes type-printer covers a sheet of paper with printed characters in bold Roman type. The transmitter has a keyboard, on which are marked letters, signs, and numbers; also a type-wheel, with the characters on its circumference, rotated by electricity. The receiver contains mechanisms for rotating another type-wheel synchronously—that is, in time—with the first; for shifting the wheel across the paper, for pressing the paper against the wheel; and for moving the paper when a fresh line is needed. These are too complicated to be described here in detail. By means of relays one transmitter may be made to work five hundred receivers. In London a single operator, controlling a keyboard in the central dispatching office, causes typewritten messages to spell themselves out simultaneously in machines distributed all over the metropolis.

The tape machine resembles that just described in many details. The main difference is that it prints on a continuous ribbon instead of on sheets.

Automatic electric printers of some kind or other are to be found in the vestibules of all the principal hotels and clubs of our large cities, and in the offices of bankers, stockbrokers, and newspaper editors. In London alone over 500 million words are printed by the receivers in a year.

HIGH-SPEED TELEGRAPHY.

At certain seasons, or when important political events are taking place, the telegraph service would become congested with news were there not some means of transmitting messages at a much greater speed than is possible by hand signalling. Fifty words a minute is about the limit speed that a good operator can maintain. By means of Wheatstone's *automatic transmitter* the rate can be increased to 400 words per minute. Paper ribbons are punched in special machines by a number of clerks with a series of holes which by their position indicate a dot or a dash. The ribbons are passed through a special transmitter, over little electric brushes, which make contact through the holes with surfaces connected to the line circuit. At the receiver end the message is printed by a Morse inker.

It has been found possible to send several messages simultaneously over a single line. To effect this a *distributer* is used to put a number of transmitters at one end of the line in communication with an equal number of receivers at the other end, fed by a second distributer keeping perfect time with the first. Instead of a signal coming as a whole to any one instrument it arrives in little bits, but these follow one another so closely as to be practically continuous. By working a number of automatic transmitters through a distributer, a thousand words or more per minute are easily dispatched over a single wire.

The Pollak Virag system employs a punched ribbon, and the receiver traces out the message in alphabetical characters on a moving strip of sensitized photographic paper. A mirror attached to a vibrating diaphragm reflects light from a lamp on to the strip, which is automatically developed and fixed in chemical baths. The method of moving the mirror so as to make the rays trace out words is extremely ingenious. Messages have been transmitted by this system at the rate of 180,000 words per hour.

CHAPTER VII.
WIRELESS TELEGRAPHY.

The transmitting apparatus—The receiving apparatus—Syntonic transmission—The advance of wireless telegraphy.

IN our last chapter we reviewed briefly some systems of sending telegraphic messages from one point of the earth's surface to another through a circuit consisting partly of an insulated wire and partly of the earth itself. The metallic portion of a long circuit, especially if it be a submarine cable, is costly to install, so that in quite the early days of telegraphy efforts were made to use the ether in the place of wire as one conductor.

When a hammer strikes an anvil the air around is violently disturbed. This disturbance spreads through the molecules of the air in much the same way as ripples spread from the splash of a stone thrown into a pond. When the sound waves reach the ear they agitate the tympanum, or drum membrane, and we "hear a noise." The hammer is here the transmitter, the air the conductor, the ear the receiver.

In wireless telegraphy we use the ether as the conductor of electrical disturbances.[13] Marconi, Slaby, Branly, Lodge, De Forest, Popoff, and others have invented apparatus for causing disturbances of the requisite kind, and for detecting their presence.

The main features of a wireless telegraphy outfit are shown in Figs. 59 and 61.

THE TRANSMITTER APPARATUS.

We will first consider the transmitting outfit (Fig. 59). It includes a battery, dispatching key, and an induction coil having its secondary circuit terminals connected with two wires, the one leading to an earth-plate, the other carried aloft on poles or suspended from a kite. In the large station at Poldhu, Cornwall, for transatlantic signalling, there are special wooden towers 215 feet high, between which the aërial wires hang. At their upper and lower ends respectively the earth and aërial wires terminate in brass balls separated by a gap. When the operator depresses the key the induction coil charges these balls and the wires attached thereto with high-tension electricity. As soon as the quantity collected exceeds the resistance of the air-gap, a discharge takes place between the balls, and the ether round the aërial wire is violently disturbed, and waves of electrical energy are propagated through it. The rapidity with which the discharges follow one another, and their travelling power, depends on the strength of the induction coil, the length of the air-gap, and the capacity of the wires.[14]

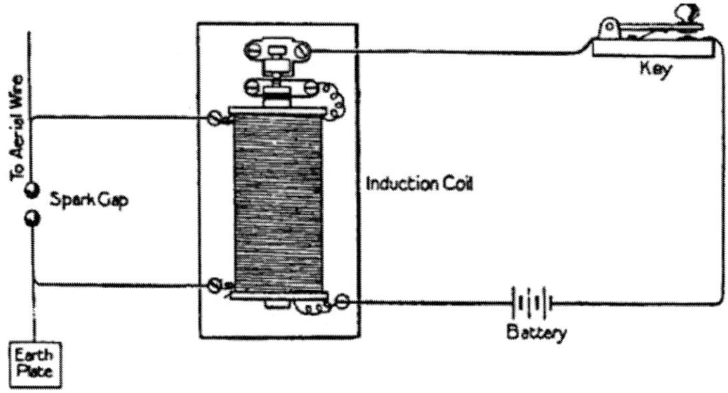

Fig. 59.—Sketch of the transmitter of a wireless telegraphy outfit.

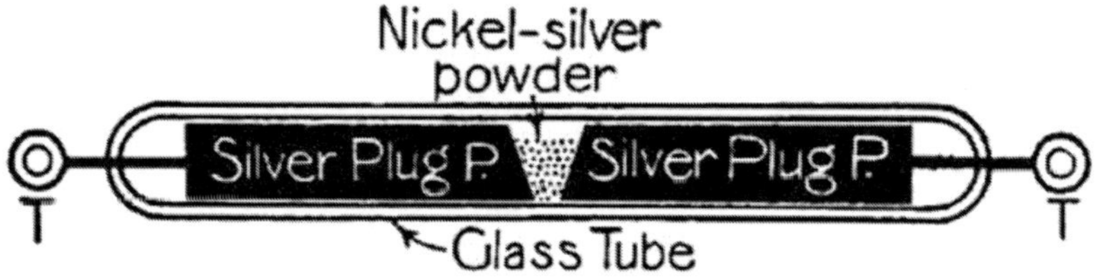

Fig. 60.—A Marconi coherer.

RECEIVING APPARATUS.

The human body is quite insensitive to these etheric waves. We cannot feel, hear, or see them. But at the receiving station there is what may be called an "electric eye." Technically it is named a *coherer*. A Marconi coherer is seen in Fig. 60. Inside a small glass tube exhausted of air are two silver plugs,P P, carrying terminals,T T, projecting through the glass at both ends. A small gap separates the plugs at the centre, and this gap is partly filled with nickel-silver powder. If the terminals of the coherer are attached to those of a battery, practically no current will pass under ordinary conditions, as the particles of nickel-silver touch each other very lightly and make a "bad contact." But if the coherer is also attached to wires leading into the earth and air, and ether waves strike those wires, at every impact the particles will cohere—that is, pack tightly together—and allow battery current to pass. The property of cohesion of small conductive bodies when influenced by Hertzian waves was first noticed in 1874 by Professor D.E. Hughes while experimenting with a telephone.

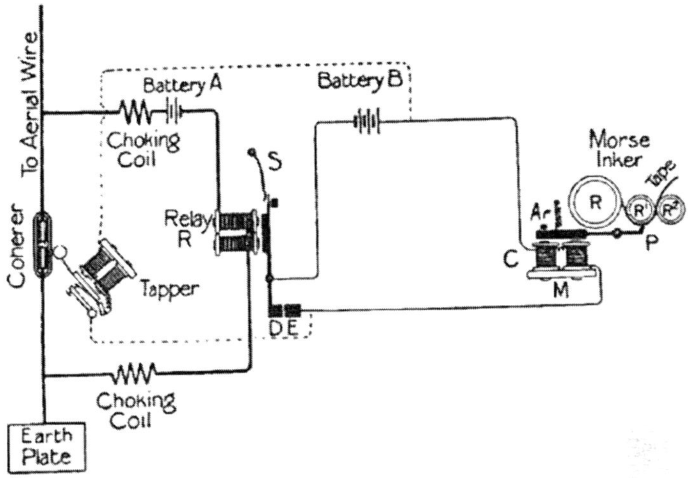

Fig. 61.—Sketch of the receiving apparatus in a wireless telegraphy outfit.

We are now in a position to examine the apparatus of which a coherer forms part (Fig. 61). First, we notice the aërial and earth wires, to which are attached other wires from batteryA. This battery circuit passes round the relay magnetRand through two choking coils, whose function is to prevent the Hertzian waves entering the battery. The relay, when energized, brings contactDagainstEand closes the circuit of batteryB, which is much more powerful than batteryA, and operates the magnetMas well as the *tapper*, which is practically an electric bell minus the gong. (The tapper circuit is indicated by the dotted lines.)

We will suppose the transmitter of a distant station to be at work. The electric waves strike the aërial wire of the receiving station, and cause the coherer to cohere and pass current. The relay is closed, and both tapper and Morse inker begin to work. The tapper keeps striking the coherer and shakes the particles loose after every cohesion. If this were not done the current ofAwould pass continuously after cohesion had once taken place. When the key of the transmitter is pressed down, the waves follow one another very quickly, and the acquired conductivity of the coherer is only momentarily destroyed by the tap of the hammer. During the impression of a dot by the Morse inker, contact is made and broken repeatedly; but as the armature of the inker is heavy and slow to move it does not vibrate in time with the relay and tapper. Therefore the Morse instrument reproduces in dots and dashes the short and long depressions of the key at the transmitting station, while the tapper works rapidly in time with the relay. The Morse inker is shown diagrammatically. While current passes throughMthe armature is pulled towards it, the endP, carrying an inked wheel, rises, and a mark is made on the tapeW, which is moved continuously being drawn forward off reelRby the clockwork—or electrically-driven rollersR1 R2.

SYNTONIC TRANSMISSION.

If a number of transmitting stations are sending out messages simultaneously, a jumble of signals would affect all the receivers round, unless some method were employed for rendering a receiver sensitive only to the waves intended to influence it. Also, if distinction were impossible, even with one transmitter in action its message might go to undesired stations.

There are various ways of "tuning" receivers and transmitters, but the principle underlying them all is analogous to that of mechanical vibration. If a weight is suspended from the end of a spiral spring, and given an upward blow, it bobs up and down a certain number of times per minute, every movement from start to finish having exactly the same duration as the rest. The resistance of the air and the internal friction of the spring gradually lessen the amplitude of the movements, and the weight finally comes to rest. Suppose that the weight scales 30 lbs., and that it naturally bobs twenty times a minute. If you now take a feather and give it a push every three seconds you can coax it into vigorous motion, assuming that every push catches it exactly on the rebound. The same effect would be produced more slowly if 6 or 9 second intervals were substituted. But if you strike it at 4, 5, or 7 second intervals it will gradually cease to oscillate, as the effect of one blow neutralizes that of another. The same phenomenon is witnessed when two tuning-forks of equal pitch are mounted near one another, and one is struck. The other soon picks up the note. But a fork of unequal pitch would remain dumb.

Now, every electrical circuit has a "natural period of oscillation" in which its electric charge vibrates. It is found possible to "tune," or "syntonize," the aërial rod or wire of a receiving station with a transmitter. A vertical wire about 200 feet in length, says Professor J.A. Fleming,[15] has a natural time period of electrical oscillation of about one-millionth of a second. Therefore if waves strike this wire a million times a second they will reinforce one another and influence the coherer; whereas a less or greater frequency will leave it practically unaffected. By adjusting the receiving circuit to the transmitter, or *vice versâ*, selective wireless telegraphy becomes possible.

ADVANCE OF WIRELESS TELEGRAPHY.

The history of wireless telegraphy may be summed up as follows:—

1842.—Professor Morse sent aërial messages across the Susquehanna River. A line containing a battery and transmitter was carried on posts along one bank and "earthed" in the river at each end. On the other bank was a second wire attached to a receiver and similarly earthed. Whenever contact was made and broken on the battery side, the receiver on the other was affected. Distance about 1 mile.

1859.—James Bowman Lindsay transmitted messages across the Tay at Glencarse in a somewhat similar way. Distance about ½ mile.

1885.—Sir William Preece signalled from Lavernock Point, near Cardiff, to Steep Holm, an island in the Bristol Channel. Distance about 5½ miles.

In all these electrical *induction* of current was employed.

1886.—Hertzian waves discovered.

1895.—Professor A. Popoff sent Hertzian wave messages over a distance of 3 miles.

1897.—Marconi signalled from the Needles Hotel, Isle of Wight, to Swanage; 17½ miles.

1901.—Messages sent at sea for 380 miles.

1901, Dec. 17.—Messages transmitted from Poldhu, Cornwall, to Hospital Point, Newfoundland; 2,099 miles.

Mr. Marconi has so perfected tuning devices that his transatlantic messages do not affect receivers placed on board ships crossing the ocean, unless they are purposely tuned. Atlantic liners now publish daily small newspapers containing the latest news, flashed through space from land stations. In the United States the De Forest and Fessenden systems are being rapidly extended to embrace the most out-of-the-way districts. Every navy of importance has adopted wireless telegraphy, which, as was proved during the Russo-Japanese War, can be of the greatest help in directing operations.

[13] Named after their first discoverer, Dr. Hertz of Carlsruhe, "Hertzian waves."

[14] For long-distance transmission powerful dynamos take the place of the induction coil and battery.

[15] "Technics," vol. ii. p. 566.

CHAPTER VIII.
THE TELEPHONE.

The Bell telephone—The Edison transmitter—The granular carbon transmitter—General arrangement of a telephone circuit—Double-line circuits—Telephone exchanges—Submarine telephony.

FOR the purposes of everyday life the telephone is even more useful than the telegraph. Telephones now connect one room of a building with another, house with house, town with town, country with country. An infinitely greater number of words pass over the telephonic circuits of the world in a year than are transmitted by telegraph operators. The telephone has become an important adjunct to the transaction of business of all sorts. Its wires penetrate everywhere. Without moving from his desk, the London citizen may hold easy converse with a Parisian, a New Yorker with a dweller in Chicago.

Wonderful as the transmission of signals over great distances is, the transmission of human speech so clearly that individual voices may be distinguished hundreds of miles away is even more so. Yet the instrument which works the miracle is essentially simple in its principles.

THE BELL TELEPHONE.

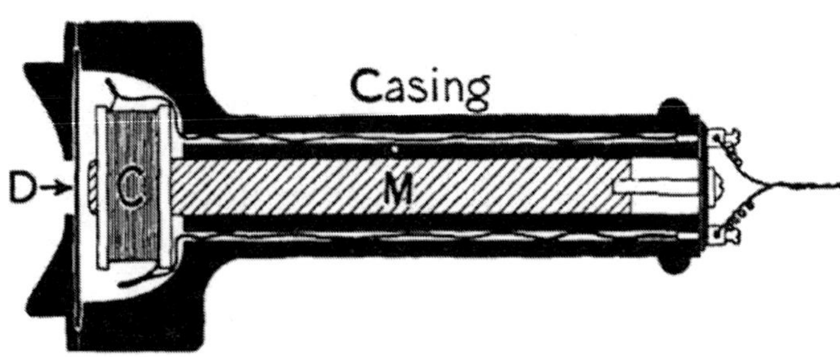

Fig. 62.—Section of a Bell telephone.

The first telephone that came into general use was that of Bell, shown in Fig. 62. In a central hole of an ebonite casing is fixed a permanent magnet,M. The casing expands at one end to accommodate a coil of insulated wire wound about one extremity of a magnet. The coil ends are attached to wires passing through small channels to terminals at the rear. A circular diaphragm,D, of very thin iron plate, clamped between the concave mouthpiece and the casing, almost touches the end of the magnet.

We will suppose that two Bell telephones, A and B, are connected up by wires, so that the wires and the coils form a complete circuit. Words are spoken into A. The air vibrations, passing through the central hole in the cover, make the diaphragm vibrate towards and away from the magnet. The distances through which the diaphragm moves have been measured, and found not to exceed in some cases more than1/10,000,000 of an inch! Its movements distort the shape of the "lines of force" emanating from the magnet, and these, cutting through the turns of the coil, induce a current in the line circuit. As the diaphragm approaches the magnet a circuit is sent in one direction; as it leaves it, in the other. Consequently speech produces rapidly alternating currents in the circuit, their duration and intensity depending on the nature of the sound.

Now consider telephone B. The currents passing through its coil increase or diminish the magnetism of the magnet, and cause it to attract its diaphragm with varying force. The vibration of the diaphragm disturbs the air in exact accordance with the vibrations of A's diaphragm, and speech is reproduced.

THE EDISON TRANSMITTER.

The Bell telephone may be used both as a transmitter and a receiver, and the permanent magnetism of the cores renders it independent of an electric battery. But currents generated by it are so minute that they cannot overcome the resistance of a long circuit; therefore a battery is now always used, and with it a special device as transmitter.

If in a circuit containing a telephone and a battery there be a loose contact, and this be shaken, the varying resistance of the contact will cause electrical currents of varying force to pass through the circuit. Edison introduced the first successful *microphone* transmitter, in which a small platinum disc connected to the diaphragm pressed with varying force against a disc of carbon, each disc forming part of the circuit. Vibrations of the diaphragm caused current to flow in a series of rapid pulsations.

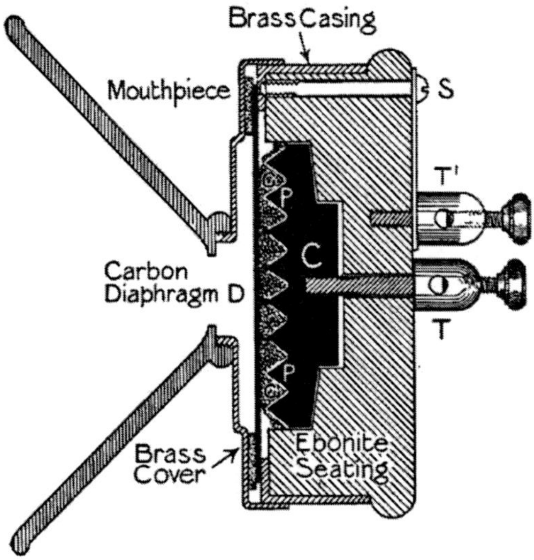

Fig. 63. — Section of a granular carbon transmitter.

THE GRANULAR CARBON TRANSMITTER.

In Fig. 63 we have a section of a microphone transmitter now very widely used. It was invented, in its original form, by an English clergyman named Hunnings. Resting in a central cavity of an ebonite seating is a carbon block,C, with a face moulded into a number of pyramidal projections,P P. The space betweenCand a carbon diaphragm,D, is packed with carbon granules,G G.Chas direct contact with line terminalT, which screws into it;DwithT1 through the brass casing, screwS, and a small plate at the back of the transmitter. Voice vibrations compressG G, and allow current to pass more freely fromDtoC. This form of microphone is very delicate, and unequalled for long-distance transmission.

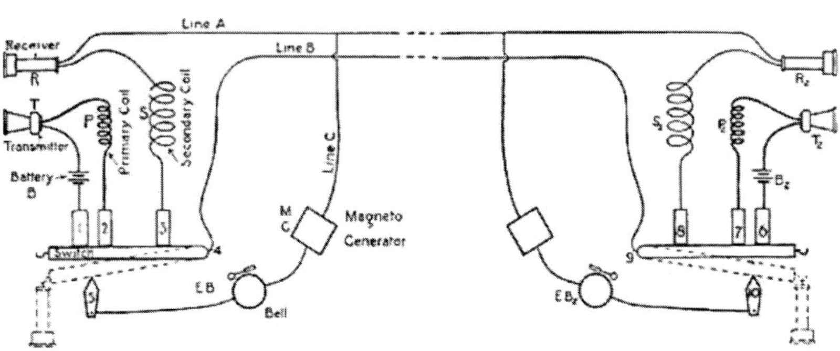

Fig. 64.—A diagrammatic representation of a telephonic circuit.

GENERAL ARRANGEMENT OF A TELEPHONE CIRCUIT.

In many forms of subscriber's instruments both receiver and transmitter are mounted on a single handle in such a way as to be conveniently placed for ear and mouth. For the sake of clearness the diagrammatic sketch of a complete installation (Fig. 64) shows them separated. The transmitters, it will be noticed, are located in battery circuits, including the primary windingsP P2 of induction coils. The transmitters are in the line circuit, which includes the secondary windingsS S2 of the coils.

We will assume that the transmitters are, in the first instance, both hung on the hooks of the metallic switches, which their weight depresses to the position indicated by the dotted lines. The handle of the magneto-generator at the left-end station is turned, and current passes through the closed circuit:—LineA,E B2, contact 10, the switch 9; lineB, 4, the other switch, contact 5, andE B. Both bells ring. Both parties now lift their receivers from the switch hooks. The switches rise against contacts 1, 2, 3 and 6, 7, 8 respectively. Both primary and both secondary circuits are now completed, while the bells are disconnected from the line wires. The pulsations set up by transmitterTin primary coilPare magnified by secondary coilSfor transmission through the line circuit, and affect both receivers. The same thing happens whenT2 is used. At the end of the conversation the receivers are hung on their hooks again, and the bell circuit is remade, ready for the next call.

A TELEPHONE EXCHANGE.

DOUBLE-LINE CIRCUITS.

The currents used in telephones pulsate very rapidly, but are very feeble. Electric disturbances caused by the proximity of telegraph or tram wires would much interfere with them if the earth were used for the return circuit. It has been found that a complete metallic circuit (two wires) is practically free from interference, though where a number of wires are hung on the same poles, speech-sounds may be faintly induced in one circuit from another. This defect is, however, minimized by crossing the wires about among themselves, so that any one line does not pass round the corresponding insulator on every pole.

TELEPHONE EXCHANGES.

In a district where a number of telephones are used the subscribers are put into connection with one another through an "exchange," to which all the wires lead. One wire of each subscriber runs to a common "earth;" the other terminates at a switchboard presided over by an operator. In an exchange used by many subscribers the terminals are distributed over a number of switchboards, each containing 80 to 100 terminals, and attended to by an operator, usually a girl.

When a subscriber wishes to be connected to another subscriber, he either turns the handle of a magneto generator, which causes a shutter to fall and expose his number at the exchange, or simply depresses a key which works a relay at the exchange and lights a tiny electric lamp. The operator, seeing the signal, connects her telephone with the subscriber's circuit and asks the number wanted. This given, she rings up the other subscriber, and connects the two circuits by means of an insulated wire cord having a spike at each end to fit the "jack" sockets of the switchboard terminals. The two subscribers are now in communication.

Fig. 65.—The headdress of an operator at a telephone exchange. The receiver is fastened over one ear, and the transmitter to the chest.

If a number on switchboard A calls for a number on switchboard C, the operator at A connects her subscriber by a jack cord to a trunk line running to C, where the operator similarly connects the trunk line with the number asked for, after ringing up the subscriber. The central exchange of one town is connected with that of another by one or more trunk lines, so that a subscriber may speak through an indefinite number of exchanges. So perfect is the modern telephone that the writer remembers on one occasion hearing the door-bell ring in a house more than a hundred miles away, with which he was at the moment in telephonic connection, though three exchanges were in the circuit.

SUBMARINE TELEPHONY.

Though telegraphic messages are transmitted easily through thousands of miles of cable,[16] submarine telephony is at present restricted to comparatively short distances. When a current passes through a cable, electricity of opposite polarity induced on the outside of the cable damps the vibration in the conductor. In the Atlantic cable, strong currents of electricity are poured periodically into one end, and though much enfeebled when they reach the other they are sufficiently strong to work a very delicate "mirror galvanometer" (invented by Lord Kelvin), which moves a reflected ray up and down a screen, the direction of the movements indicating a dot or a dash. Reversible currents are used in transmarine telegraphy. The galvanometer is affected like the coils and small magnet in Wheatstone's needle instrument.

Telephonic currents are too feeble to penetrate many miles of cable. There is telephonic communication between England and France, and England and Ireland. But transatlantic telephony is still a thing of the future. It is hoped, however, that by inserting induction coils at intervals along the cables the currents may be "stepped up" from point to point, and so get across. Turning to Fig. 64, we may supposeSto be on shore at the English end, andS2 to be the *primary* winding of an induction coil a hundred miles away in the sea, which magnifies the enfeebled vibrations for a journey toS3, where they are again revived; and so on, till the New World is reached. The difficulty is to devise induction coils of great power though of small size. Yet science advances nowadays so fast that we may live to hear words spoken at the Antipodes.

[16] In 1896 the late Li Hung Chang sent a cablegram from China to England (12,608 miles), and received a reply, in *seven minutes.*

CHAPTER IX.
DYNAMOS AND ELECTRIC MOTORS.

A simple dynamo—Continuous-current dynamos—Multipolar dynamos—Exciting the field magnets—Alternating current dynamos—The transmission of power—The electric motor—Electric lighting—The incandescent lamp—Arc lamps—"Series" and "parallel" arrangement of lamps—Current for electric lamps—Electroplating.

IN previous chapters we have incidentally referred to the conversion of mechanical work into electrical energy. In this we shall examine how it is done—how the silently spinning dynamo develops power, and why the motor spins when current is passed through it.

We must begin by returning to our first electrical diagram (Fig. 50), and calling to mind the invisible "lines of force" which permeate the ether in the immediate neighbourhood of a magnet's poles, called the *magnetic field* of the magnet.

Many years ago (1831) the great Michael Faraday discovered that if a loop of wire were moved up and down between the poles of an electro-magnet (Fig. 66) a current was induced in the loop, its direction depending upon that in which the loop was moved. The energy required to cut the lines of force passed in some mysterious way into the wire. Why this is so we cannot say, but, taking advantage of the fact, electricians have gradually developed the enormous machines which now send vehicles spinning over metal tracks, light our streets and houses, and supply energy to innumerable factories.

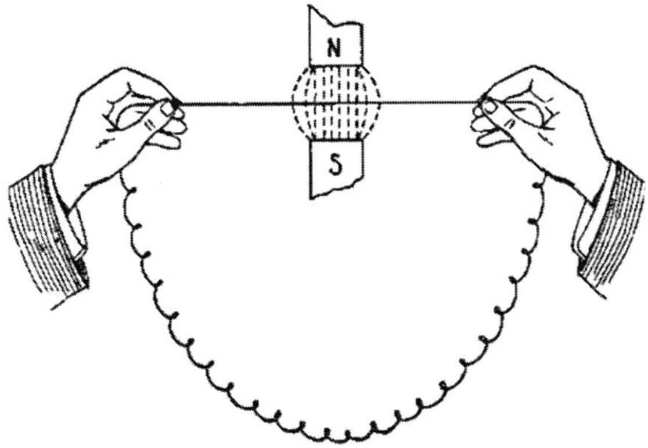

Fig. 66.

The strength of the current induced in a circuit cutting the lines of force of a magnet is called its pressure, voltage, or electro-motive force (expressed shortly E.M.F.). It may be compared with the pounds-to-the-square-inch of steam. In order to produce an E.M.F. of one volt it is calculated that 100,000,000 lines of force must be cut every second.

The voltage depends on three things:—(1.) The *strength* of the magnet: the stronger it is, the greater the number of lines of force coming from it. (2.) The *length* of the conductor cutting the lines of force: the longer it is, the more lines it will cut. (3.) The *speed* at which the conductor moves: the faster it travels, the more lines it will cut in a given time. It follows that a powerful dynamo, or mechanical producer of current, must have strong magnets and a long conductor; and the latter must be moved at a high speed across the lines of force.

A SIMPLE DYNAMO.

In Fig. 67 we have the simplest possible form of dynamo—a single turn of wire, *w x y z*, mounted on a spindle, and having one end attached to an insulated ringC, the other to an insulated ringC1. Two small brushes,B B1, of wire gauze or carbon, rubbing continuously against these collecting rings, connect them with a wire which completes the circuit. The armature, as the revolving coil is called, is mounted between the poles of a magnet, where the lines of force are thickest. These lines are *supposed* to stream from the N. to the S. pole.

In Fig. 67 the armature has reached a position in which *y z* and *w x* are cutting no, or very few, lines of force, as they move practically parallel to the lines. This is called the *zero* position.

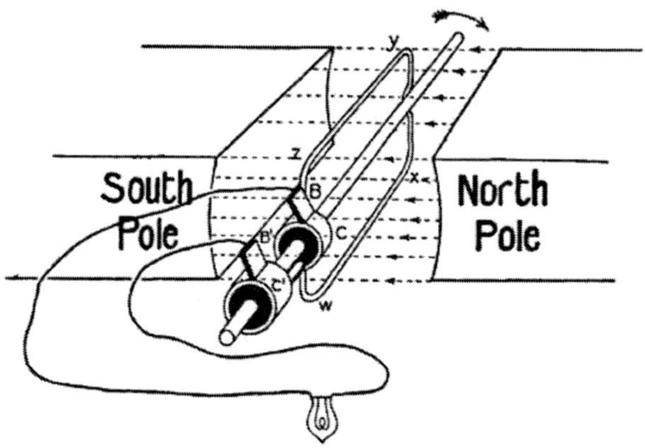

Fig. 67.

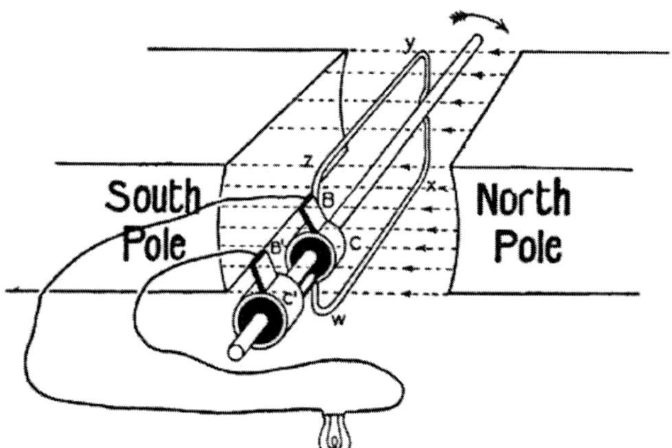

Fig. 68.

In Fig. 68 the armature, moving at right angles to the lines of force, cuts a maximum number in a given time, and the current induced in the coil is therefore now most intense. Here we must stop a moment to consider how to decide in which direction the current flows. The armature is revolving in a clockwise direction, and *y z*, therefore, is moving downwards. Now, suppose that you rest your *left* hand on the N. pole of the magnet so that the arm lies in a line with the magnet. Point your forefinger towards the S. pole. It will indicate the *direction of the lines of force*. Bend your other three fingers downwards over the edge of the N. pole. They will indicate the *direction in which the conductor is moving* across the magnetic field. Stick out the thumb at right angles to the forefinger. It points in the direction in which the *induced* current is moving through the nearer half of the coil. Therefore lines of force, conductor, and induced current travel in planes which, like the top and two adjacent sides of a box, are at right angles to one another.

While current travels from *z* to *y*—that is, *from* the ringC1 to *y*—it also travels from *x* to *w*, because *w x* rises while *y z* descends. So that a current circulates through the coil and the exterior part of the circuit, including the lamp. After *z y* has passed the lowest possible point of the circle it begins to ascend, *w x* to descend. The direction of the current is therefore reversed; and as the change is repeated every half-revolution this form of dynamo is called an *alternator* or creator of alternating currents. A well-known type of alternator is the magneto machine which sends shocks through any one who completes the external circuit by holding the brass handles connected by wires to the brushes. The faster the handle of the machine is turned the more frequent is the alternation, and the stronger the current.

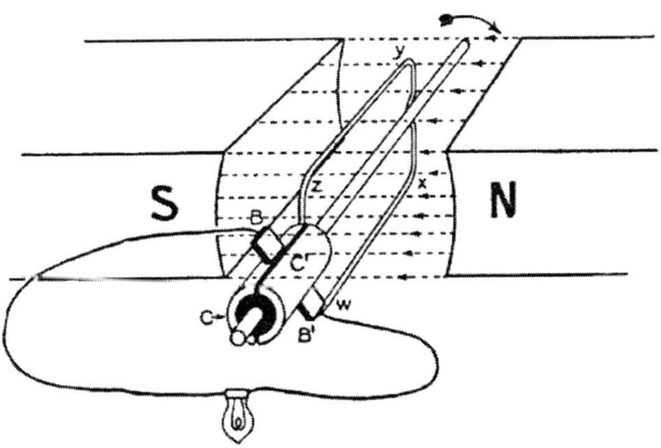

Fig. 69.

CONTINUOUS-CURRENT DYNAMOS.

An alternating current is not so convenient for some purposes as a continuous current. It is therefore sometimes desirable (even necessary) to convert the alternating into a uni-directional or continuous current. How this is done is shown in Figs. 69 and 70. In place of the two collecting ringsC C1, we now have a single ring split longitudinally into two portions, one of which is connected to each end of the coil *w x y z*. In Fig. 69 brushBhas just passed the gap on to segmentC, brushB1 on to segmentC1. For half a revolution these remain respectively in contact; then, just as *y z* begins to rise and *w x* to descend, the brushes cross the gaps again and exchange segments, so that the current is perpetually flowing one way through the circuit. The effect of the commutator[17] is, in fact, equivalent to transposing the brushes of the collecting rings of the alternator every time the coil reaches a zero position.

Figs. 71 and 72 give end views in section of the coil and the commutator, with the coil in the position of minimum and maximum efficiency. The arrow denotes the direction of movement; the double dotted lines the commutator end of the revolving coil.

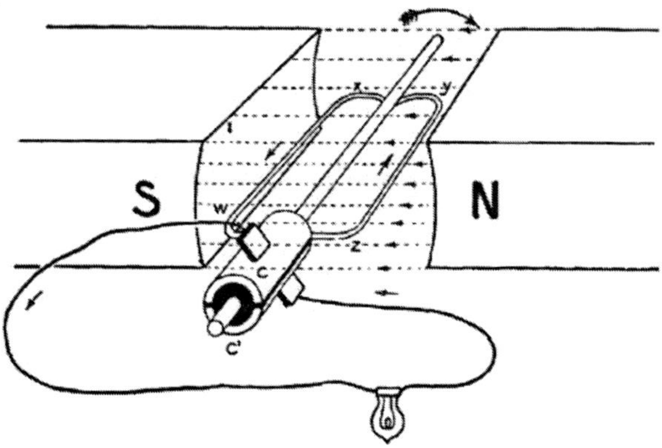

Fig. 70.

PRACTICAL CONTINUOUS-CURRENT DYNAMOS.

The electrical output of our simple dynamo would be increased if, instead of a single turn of wire, we used a coil of many turns. A further improvement would result from mounting on the shaft, inside the coil, a core or drum of iron, to entice the lines of force within reach of the revolving coil. It is evident that any lines which pass through the air outside the circle described by the coil cannot be cut, and are wasted.

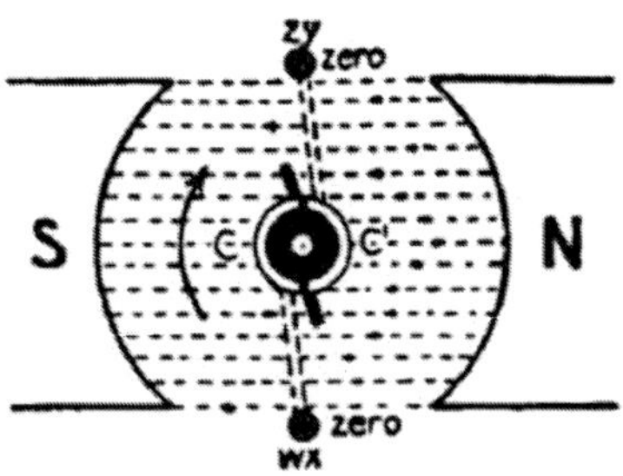

Fig. 71.

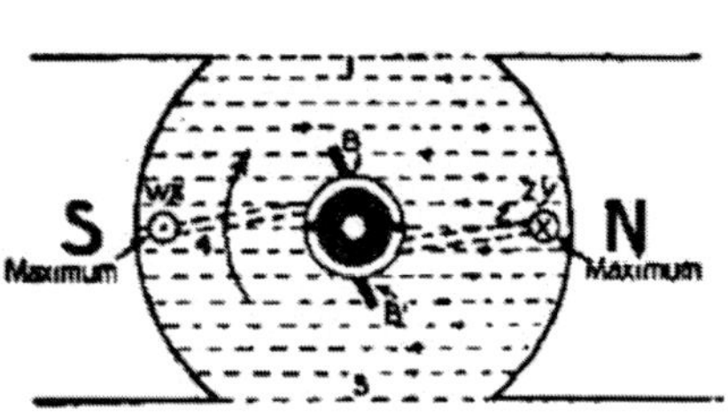

Fig. 72.

The core is not a solid mass of iron, but built up of a number of very thin iron discs threaded on the shaft and insulated from one another to prevent electric eddies, which would interfere with the induced current in the conductor.[18] Sometimes there are openings through the core from end to end to ventilate and cool it.

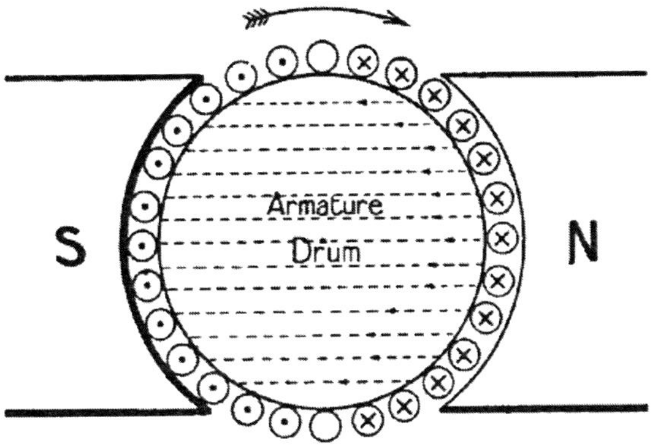

Fig. 73.

We have already noticed that in the case of a single coil the current rises and falls in a series of pulsations. Such a form of armature would be unsuitable for large dynamos, which accordingly have a number of coils wound over their drums, at equal distances round the circumference, and a commutator divided into an equal number of segments. The subject of drum winding is too complicated for brief treatment, and we must therefore be content with noticing that the coils are so connected to their respective commutator segments and to one another that they mutually assist one another. A glance at Fig. 73 will help to explain this. Here we have in section a number of conductors on the right of the drum (marked with a cross to show that current is moving, as it were, into the page), connected with conductors on the left (marked with a dot to signify current coming out of the page). If the "crossed" and "dotted" conductors were respectively the "up" and "down" turns of a single coil terminating in a simple split commutator (Fig. 69), when the coil had been revolved through an angle of 90° some of the up turns would be ascending and some descending, so that conflicting currents would arise. Yet we want to utilize the whole surface of the drum; and by winding a number of coils in the manner hinted at, each coil, as it passes the zero point, top or bottom, at once generates a current in the desired direction and reinforces that in all the other turns of its own and of other coils on the same side of a line drawn vertically through the centre. There is thus practically no fluctuation in the pressure of the current generated.

The action of single and multiple coil windings may be compared to that of single and multiple pumps. Water is ejected by a single pump in gulps; whereas the flow from a pipe fed by several pumps arranged to deliver consecutively is much more constant.

MULTIPOLAR DYNAMOS.

Hitherto we have considered the magnetic field produced by one bi-polar magnet only. Large dynamos have four, six, eight, or more field magnets set inside a casing, from which their cores project towards the armature so as almost to touch it (Fig. 74). The magnet coils are wound to give N. and S. poles alternately at their armature ends round the field; and the lines of force from each N. pole stream each way to the two adjacent S. poles across the path of the armature coils. In dynamos of this kind several pairs of collecting brushes pick current off the commutator at equidistant points on its circumference.

Fig. 74.—A Holmes continuous current dynamo:A, armature; C, commutator; M, field magnets.

EXCITING THE FIELD MAGNETS.

Until current passes through the field magnet coils, no magnetic field can be created. How are the coils supplied with current? A dynamo, starting for the first time, is excited by a current from an outside source; but when it has once begun to generate current it feeds its magnets itself, and ever afterwards will be self-exciting,[19] owing to the residual magnetism left in the magnet cores.

Fig. 75.—Partly finished commutator.

Look carefully at Figs. 77 and 78. In the first of these you will observe that part of the wire forming the external circuit is wound round the arms of the field magnet. This is called a *series* winding. In this case *all* the current generated helps to excite the dynamo. At the start the residual magnetism of the magnet cores gives a weak field. The armature coils cut this and pass a current through the circuit. The magnets are further excited, and the field becomes stronger; and so on till the dynamo is developing full power. Series winding is used where the current in the external circuit is required to be very constant.

Fig. 76.—The brushes of a Holmes dynamo.

Fig. 78 shows another method of winding—the *shunt*. Most of the current generated passes through the external circuit 2, 2; but a part is switched through a separate winding for the magnets, denoted by the fine wire 1, 1. Here the strength of the magnetism does not vary directly with the current, as only a small part of the current serves the magnets. The shunt winding is therefore used where the voltage (or pressure) must be constant.

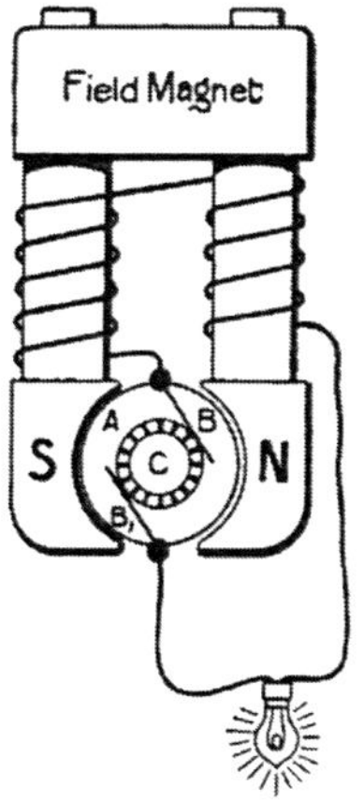

Fig. 77.—Sketch showing a "series" winding.

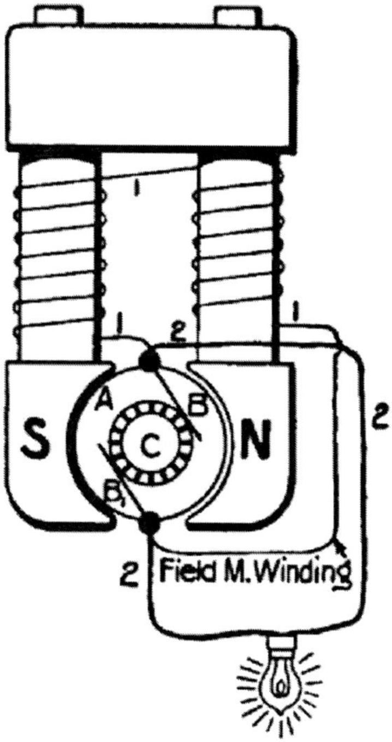

Fig. 78.—"Shunt" winding.

A third method is a combination of the two already named. A winding of fine wire passes from brush to brush round the magnets; and there is also a series winding as in Fig. 77. This compound method is adapted more especially for electric traction.

ALTERNATING DYNAMOS.

These have their field magnets excited by a separate continuous current dynamo of small size. The field magnets usually revolve inside a fixed armature (the reverse of the arrangement in a direct-current generator); or there may be a fixed central armature and field magnets revolving outside it. This latter arrangement is found in the great power stations at Niagara Falls, where the enormous field-rings are mounted on the top ends of vertical shafts, driven by water-turbines at the bottom of pits 178 feet deep, down which water is led to the turbines through great pipes, or penstocks. The weight of each shaft and the field-ring attached totals about thirty-five tons. This mass revolves 250 times a minute, and 5,000 horse power is constantly developed by the dynamo. Similar dynamos of 10,000 horse power each have been installed on the Canadian side of the Falls.

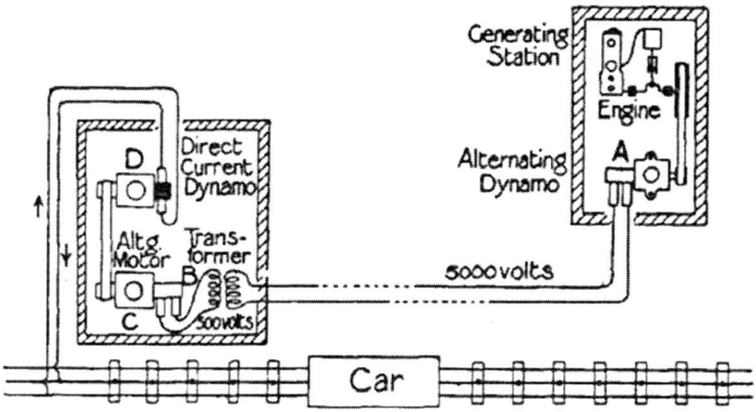

Fig. 79.

TRANSMISSION OF POWER.

Alternating current is used where power has to be transmitted for long distances, because such a current can be intensified, or stepped up, by a transformer somewhat similar in principle to a Ruhmkorff coil *minus* a contact-breaker. A typical example of transformation is seen in Fig. 79. Alternating current of 5,000 volts pressure is produced in the generating station and sent through conductors to a distant station, where a transformer,B, reduces the pressure to 500 volts to drive an alternating motor,C, which in turn operates a direct current dynamo,D. This dynamo has its terminal connected with the insulated or "live" rail of an electric railway, and its – terminal with the wheel rails, which are metallically united at the joints to act as a "return." On its way from the live rail to the return the current passes through the motors. In the case of trams the conductor is either a cable carried overhead on standards, from which it passes to the motor through a trolley arm, or a rail laid underground in a conduit between the rails. In the top of the conduit is a slit through which an arm carrying a contact shoe on the end projects from the car. The shoe rubs continuously on the live rail as the car moves.

To return for a moment to the question of transformation of current. "Why," it may be asked, "should we not send low-pressure *direct* current to a distant station straight from the dynamo, instead of altering its nature and pressure? Or, at any rate, why not use high-pressure direct current, and transform *that*?" The answer is, that to transmit a large amount of electrical energy at low pressure (or voltage) would necessitate large volume (or *amperage*) and a big and expensive copper conductor to carry it. High-pressure direct current is not easily generated, since the sparking at the collecting brushes as they pass over the commutator segments gives trouble. So engineers prefer high-pressure alternating current, which is easily produced, and can be sent through a small and inexpensive conductor with little loss. Also its voltage can be transformed by apparatus having no revolving parts.

THE ELECTRIC MOTOR.

Anybody who understands the dynamo will also be able to understand the electric motor, which is merely a reversed dynamo.

Imagine in Fig. 70 a dynamo taking the place of the lamp and passing current through the brushes and commutator into the coil *w x y z*. Now, any coil through which current passes becomes a magnet with N. and S. poles at either end. (In Fig. 70 we will assume that the N. pole is below and the S. pole above the coil.) The coil poles therefore try to seek the contrary poles of the permanent magnet, and the coil revolves until its S. pole faces the N. of the magnet, and *vice versâ*. The lines of force of the coil and the magnet are now parallel. But the momentum of revolution carries the coil on, and suddenly the commutator reverses its polarity, and a further half-revolution takes place. Then comes a further reversal, and so on *ad infinitum*. The rotation of the motor is therefore merely a question of repulsion and attraction of like and unlike poles. An ordinary compass needle may be converted into a tiny motor by presenting the N. and S. poles of a magnet to its S. and N. poles alternately every half-revolution.

In construction and winding a motor is practically the same as a dynamo. In fact, either machine can perform either function, though perhaps not equally well adapted for both. Motors may be run with direct or alternating current, according to their construction.

On electric cars the motor is generally suspended from the wheel truck, and a small pinion on the armature shaft gears with a large pinion on a wheel axle. One great advantage of electric traction is that every vehicle of a train can carry its own motor, so that the whole weight of the train may be used to get a grip on the rails when starting. Where a single steam locomotive is used, the adhesion of its driving-wheels only is available for overcoming the inertia of the load; and the whole strain of starting is thrown on to the foremost couplings. Other advantages may be summed up as follows:—(1) Ease of starting and rapid acceleration; (2) absence of waste of energy (in the shape of burning fuel) when the vehicles are at rest; (3) absence of smoke and smell.

ELECTRIC LIGHTING.

Dynamos are used to generate current for two main purposes—(1) To supply power to motors of all kinds; (2) to light our houses, factories, and streets. In private houses and theatres incandescent lamps are generally used; in the open air, in shops, and in larger buildings, such as railway stations, the arc lamp is more often found.

INCANDESCENT LAMP.

If you take a piece of very fine iron wire and lay it across the terminals of an accumulator, it becomes white hot and melts, owing to the heat generated by its resistance to the current. A piece of fine platinum wire would become white hot without melting, and would give out an intense light. Here we have the principle of the glow or incandescent lamp—namely, the interposition in an electric circuit of a conductor which at once offers a high resistance to the current, but is not destroyed by the resulting heat.

In Fig. 80 is shown a fan propelling liquid constantly through a pipe. Let us assume that the liquid is one which develops great friction on the inside of the pipe. At the contraction, where the speed of travel is much greater than elsewhere in the circuit, most heat will be produced.

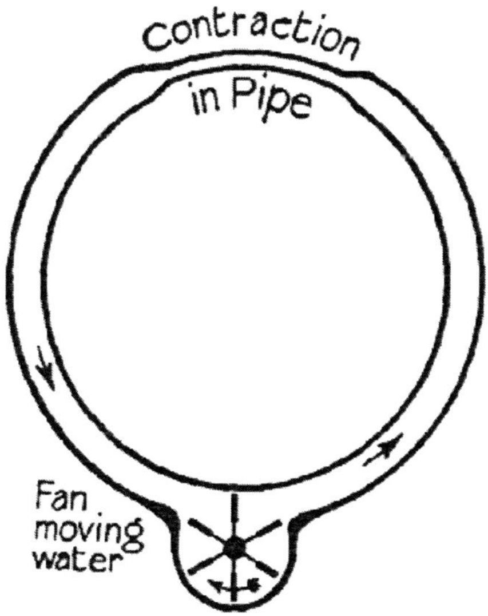

Fig. 80.—Diagram to show circulation of water through a pipe.

In quite the early days of the glow-lamp platinum wire was found to be unreliable as regards melting, and filaments of carbon are now used. To prevent the wasting away of the carbon by combination with oxygen the filament is enclosed in a glass bulb from which practically all air has been sucked by a mercury pump before sealing.

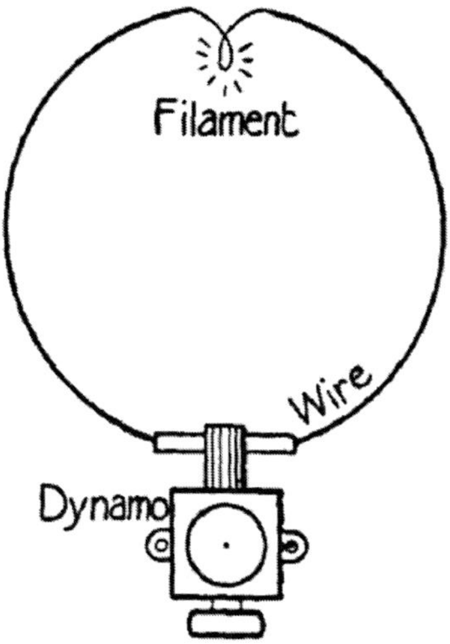

Fig. 81.—The electrical counterpart of Fig. 80. The filament takes the place of the contraction in the pipe.

The manufacture of glow-lamps is now an important industry. One brand of lamp[20] is made as follows:—First, cotton-wool is dissolved in chloride of zinc, and forms a treacly solution, which is squirted through a fine nozzle into a settling solution which hardens it and makes it coil up like a very fine violin string. After being washed and dried, it is wound on a plumbago rod and baked in a furnace until only the carbon element remains. This is the filament in the rough. It is next removed from the rod and tipped with two short pieces of fine platinum wire. To make the junction electrically perfect the filament is plunged in benzine and heated to whiteness by the passage of a strong current, which deposits the carbon of the benzine on the joints. The filament is now placed under the glass receiver of an air-pump, the air is exhausted, hydro-carbon vapour is introduced, and the filament has a current passed through it to make it white hot. Carbon from the vapour is deposited all over the filament until the required electrical resistance is attained. The filament is now ready for enclosure in the bulb. When the bulb has been exhausted and sealed, the lamp is tested, and, if passed, goes to the finishing department, where the two platinum wires (projecting through the glass) are soldered to a couple of brass plates, which make contact with two terminals in a lamp socket. Finally, brass caps are affixed with a special water-tight and hard cement.

ARC LAMPS.

In *arc* lighting, instead of a contraction at a point in the circuit, there is an actual break of very small extent. Suppose that to the ends of the wires leading from a dynamo's terminals we attach two carbon rods, and touch the end of the rods together. The tips become white hot, and if they are separated slightly, atoms of incandescent carbon leap from the positive to the negative rod in a continuous and intensely luminous stream, which is called an *arc* because the path of the particles is curved. No arc would be formed unless the carbons were first touched to start incandescence. If they are separated too far for the strength of the current to bridge the gap the light will flicker or go out. The arc lamp is therefore provided with a mechanism which, when the current is cut off, causes the carbons to fall together, gradually separates them when it is turned on, and keeps them apart. The principle employed is the effort of a coil through which a current passes to draw an iron rod into its centre. Some of the current feeding the lamp is shunted through a coil, into which projects one end of an iron bar connected with one carbon point. A spring normally presses the points together when no current flows. As soon as current circulates through the coil the bar is drawn upwards against the spring.

SERIES AND PARALLEL ARRANGEMENT OF LAMPS.

When current passes from one lamp to another, as in Fig. 82, the lamps are said to be in *series*. Should one lamp fail, all in the circuit would go out. But where arc lamps are thus arranged a special mechanism on each lamp "short circuits" it in case of failure, so that current may pass uninterruptedly to the next.

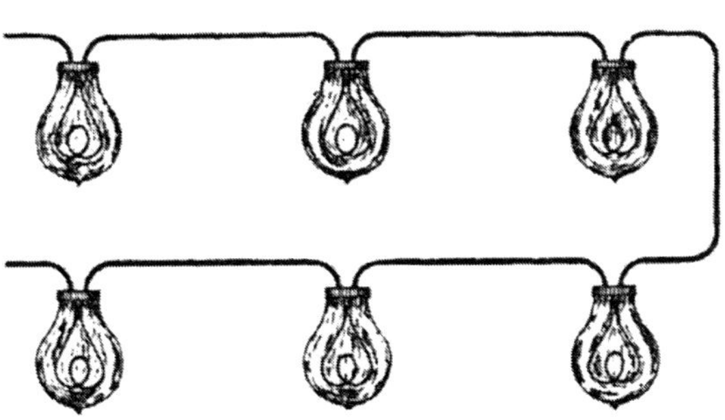

Series

Fig. 82.—Incandescent lamps connected in "series."

Fig. 83 shows a number of lamps set *in parallel*. One terminal of each is attached to the positive conductor, the other to the negative conductor. Each lamp therefore forms an independent bridge, and does not affect the efficiency of the rest. *Parallel series* signifies a combination of the two systems, and would be illustrated if, in Fig. 83, two or more lamps were connected in series groups from one conductor to the other. This arrangement is often used in arc lighting.

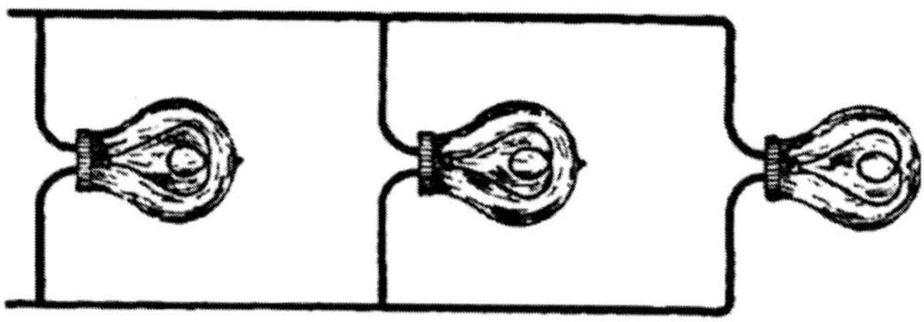

Fig. 83.—Incandescent lamps connected in "parallel."

CURRENT FOR ELECTRIC LAMPS.

This may be either direct or alternating. The former is commonly used for arc lamps, the latter for incandescent, as it is easily stepped-down from the high-pressure mains for use in a house. Glow-lamps usually take current of 110 or 250 volts pressure.

In arc lamps fed with direct current the tip of the positive carbon has a bowl-shaped depression worn in it, while the negative tip is pointed. Most of the illumination comes from the inner surface of the bowl, and the positive carbon is therefore placed uppermost to throw the light downwards. An alternating current, of course, affects both carbons in the same manner, and there is no bowl.

The carbons need frequent renewal. A powerful lamp uses about 70 feet of rod in 1,000 hours if the arc is exposed to the air. Some lamps have partly enclosed arcs—that is, are surrounded by globes perforated by a single small hole, which renders combustion very slow, though preventing a vacuum.

ELECTROPLATING.

Electroplating is the art of coating metals with metals by means of electricity. Silver, copper, and nickel are the metals most generally deposited. The article to be coated is suspended in a chemical solution of the metal to be deposited. Fig. 84 shows a very simple plating outfit.Ais a battery;Ba vessel containing, say, an acidulated solution of sulphate of copper. A spoon,S, hanging in this from a glass rod,R, is connected with the zinc or negative element,Z, of the battery, and a plate of copper,P, with the positive element,C. Current flows in the direction shown by the arrows, fromZtoC,CtoP,PtoS,StoZ. The copper deposited from the solution on the spoon is replaced by gradual dissolution of the plate, so that the latter serves a double purpose.

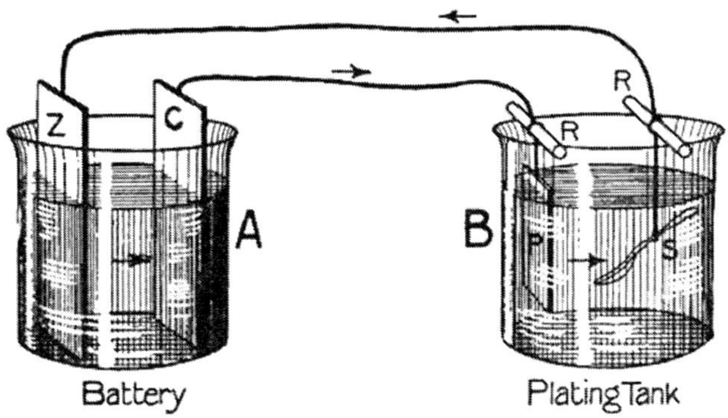

Fig. 84.—An electroplating outfit.

In silver plating, P is of silver, and the solution one of cyanide of potassium and silver salts. Where nickel or silver has to be deposited on iron, the article is often given a preliminary coating of copper, as iron does not make a good junction with either of the first two metals, but has an affinity for copper.

[17] From the Latin *commuto,* "I exchange."

[18] Only the "drum" type of armature is treated here.

[19] This refers to continuous-current dynamos only.

[20] The Robertson.

CHAPTER X.
RAILWAY BRAKES.

The Vacuum Automatic brake—The Westinghouse air-brake.

IN the early days of the railway, the pulling up of a train necessitated the shutting off of steam while the stopping-place was still a great distance away. The train gradually lost its velocity, the process being hastened to a comparatively small degree by the screw-down brakes on the engine and guard's van. The goods train of to-day in many cases still observes this practice, long obsolete in passenger traffic.

An advance was made when a chain, running along the entire length of the train, was arranged so as to pull on subsidiary chains branching off under each carriage and operating levers connected with brake blocks pressing on every pair of wheels. The guard strained the main chain by means of a wheel gear in his van. This system was, however, radically defective, since, if any one branch chain was shorter than the rest, it alone would get the strain. Furthermore, it is obvious that the snapping of the main chain would render the whole arrangement powerless. Accordingly, brakes operated by steam were tried. Under every carriage was placed a cylinder, in connection with a main steam-pipe running under the train. When the engineer wished to apply the brakes, he turned high-pressure steam into the train pipe, and the steam, passing into the brake cylinders, drove out in each a piston operating the brake gear. Unfortunately, the steam, during its passage along the pipe, was condensed, and in cold weather failed to reach the rear carriages. Water formed in the pipes, and this was liable to freeze. If the train parted accidentally, the apparatus of course broke down.

Hydraulic brakes have been tried; but these are open to several objections; and railway engineers now make use of air-pressure as the most suitable form of power. Whatever air system be adopted, experience has shown that three features are essential:—(1.) The brakes must be kept "off" artificially. (2.) In case of the train parting accidentally, the brakes must be applied automatically, and quickly bring all the vehicles of the train to a standstill. (3.)

It must be possible to apply the brakes with greater or less force, according to the needs of the case.

At the present day one or other of two systems is used on practically all automatically-braked cars and coaches. These are known as—(1) The *vacuum automatic*, using the pressure of the atmosphere on a piston from the other side of which air has been mechanically exhausted; and (2) the *Westinghouse automatic*, using compressed air. The action of these brakes will now be explained as simply as possible.

THE VACUUM AUTOMATIC BRAKE.

Under each carriage is a vacuum chamber (Fig. 85) riding on trunnions,E E, so that it may swing a little when the brakes are applied. Inside the chamber is a cylinder, the piston of which is rendered air-tight by a rubber ring rolling between it and the cylinder walls. The piston rod works through an air-tight stuffing-box in the bottom of the casing, and when it rises operates the brake rods. It is obvious that if air is exhausted from both sides of the piston at once, the piston will sink by reason of its own weight and that of its attachments. If air is now admitted below the piston, the latter will be pushed upwards with a maximum pressure of 15 lbs. to the square inch. The ball-valve ensures that while air can be sucked from *both* sides of the piston, it can be admitted to the lower side only.

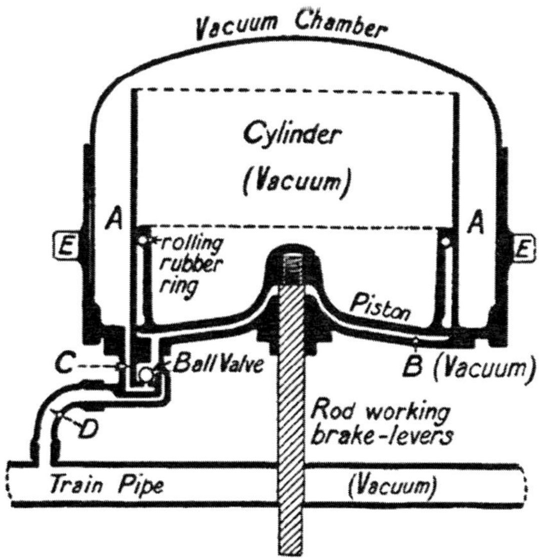

Fig. 85.—Vacuum brake "off."

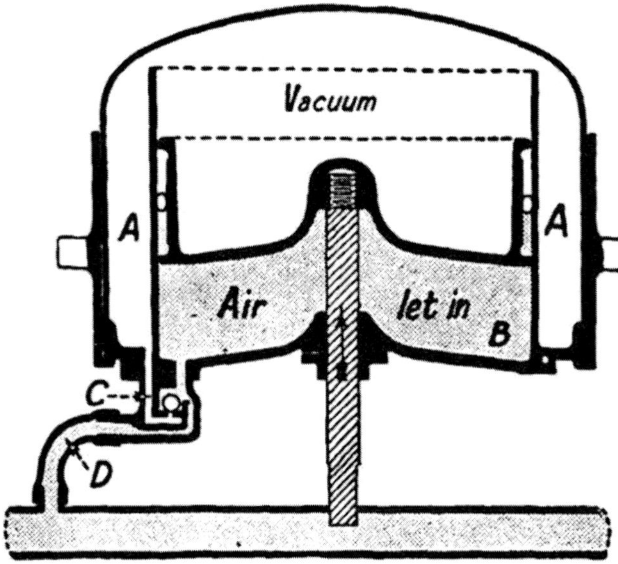

Fig. 86.—Vacuum brake "on."

Let us imagine that a train has been standing in a siding, and that air has gradually filled the vacuum chamber by leakage. The engine is coupled on, and the driver at once turns on the steam ejector,[21] which sucks all the air out of the pipes and chambers throughout the train. The air is sucked directly from the under side of the piston through pipeD; and from the spaceA Aand the cylinder (open at the top) through the channelC, lifting the ball, which, as soon as exhaustion is complete, or when the pressure on both sides of the piston is equal, falls back on its seat. On air being admitted to the train pipe, it rushes throughDand into the spaceB(Fig. 86) below the piston, but is unable to pass the ball, so that a strong upward pressure is exerted on the piston, and the brakes go on. To throw them off, the space below the piston must be exhausted. This is to be noted: If there is a leak, as in the case of the train parting, *the brakes go on at once*, since the vacuum below the piston is automatically broken.

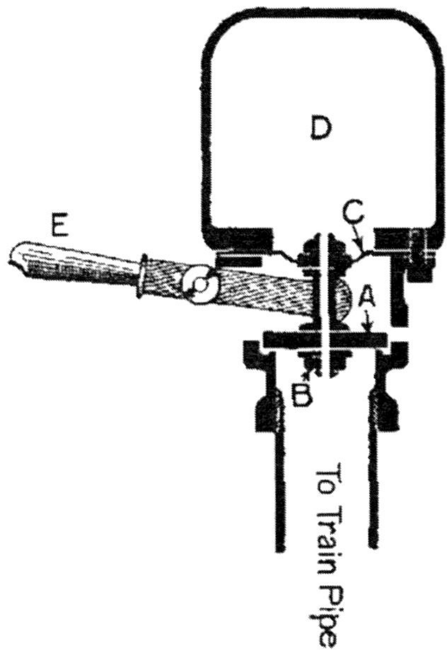

Fig. 87.—Guard's valve for applying the Vacuum brake.

For ordinary stops the vacuum is only partially broken—that is, an air-pressure of but from 5 to 10 lbs. per square inch is admitted. For emergency stops full atmospheric pressure is used. In this case it is advisable that air should enter at *both* ends of the train; so in the guard's van there is installed an ingenious automatic valve, which can at any time be opened by the guard pressing down a lever, but which opens of itself when the train-pipe vacuum is rapidly destroyed. Fig. 87 shows this device in section. Seated on the top of an upright pipe is a valve, *A*, connected by a bolt,B, to an elastic diaphragm,C, sealing the bottom of the chamberD. The boltBhas a very small hole bored through it from end to end. When the vacuum is broken slowly, the pressure falls inDas fast as in the pipe; but a sudden inrush of air causes the valveAto be pulled off its seat by the diaphragmC, as the vacuum inDhas not been broken to any appreciable extent. Air then rushes into the train pipe through the valve. It is thus evident that the driver controls this valve as effectively as if it were on the engine. These "emergency" valves are sometimes fitted to every vehicle of a train.

When a carriage is slipped, taps on each side of the coupling joint of the train pipe are turned off by the guard in the "slip;" and when he wishes to stop he merely depresses the leverE, gradually opening the valve. Under the van is an auxiliary vacuum chamber, from which the air is exhausted by the train pipe. If the guard, after the slip has parted from the train, finds that he has applied his brakes too hard, he can put this chamber into communication with the brake cylinder, and restore the vacuum sufficiently to pull the brakes off again.

When a train has come to rest, the brakes must be sucked off by the ejector. Until this has been done the train cannot be moved, so that it is impossible for it to leave the station unprepared to make a sudden stop if necessary.

THE WESTINGHOUSE AIR-BRAKE.

This system is somewhat more complicated than the vacuum, though equally reliable and powerful. Owing to the complexity of certain parts, such as the steam air-pump and the triple-valve, it is impossible to explain the system in detail; we therefore have recourse to simple diagrammatic sketches, which will help to make clear the general principles employed.

The air-brake, as first evolved by Mr. George Westinghouse, was a very simple affair—an air-pump and reservoir on the engine; a long pipe running along the train; and a cylinder under every vehicle to work the brakes. To stop the train, the high-pressure air collected in the reservoir was turned into the train pipe to force out the pistons in the coach cylinders, connected to it by short branch pipes. One defect of this "straight" system was that the brakes at the rear of a long train did not come into action until a considerable time after the driver turned on the air; and since, when danger is imminent, a very few seconds are of great importance, this slowness of operation was a serious fault. Also, it was found that the brakes on coaches near the engine went on long before those more distant, so that during a quick stop there was a danger of the forward coaches being bumped by those behind. It goes without saying that any coaches which might break loose were uncontrollable. Mr. Westinghouse therefore patented his *automatic* brake, now so largely used all over the world. The brake ensures practically instantaneous and simultaneous action on all the vehicles of *a train of any length.*

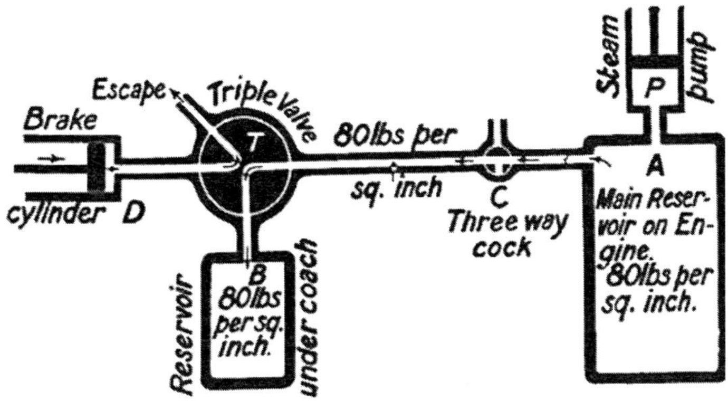

Fig. 88.—Diagrammatic sketch of the details of the Westinghouse air-brake. Brake "off."

The principle of the brake will be gathered from Figs. 88 and 89.Pis a steam-driven air-pump on the engine, which compresses air into a reservoir,A, situated below the engine or tender, and maintains a pressure of from 80 to 90 lbs. per square inch. A three-way cock,C, puts the train pipe into communication withAor the open air at the wish of the driver. Under each coach is a triple-valve,T, an auxiliary reservoir,B, and a brake cylinder,D. The triple-valve is the most noteworthy feature of the whole system. The reader must remember that the valve shown in the section is *only diagrammatic*.

Now for the operation of the brake. When the engine is coupled to the train, the compressed air in the main reservoir is turned into the train pipe, from which it passes through the triple-valve into the auxiliary reservoir, and fills it till it has a pressure of, say, 80 lbs. per square inch. Until the brakes are required, the pressure in the train pipe must be maintained. If accidentally, or purposely (by turning the cockCto the position shown in Fig. 89), the train-pipe pressure is reduced, the triple-valve at once shifts, puttingBin connection with the brake cylinderD, and cutting off the connection betweenDand the air, and the brakes go on. To get them off, the pressure in the train pipe must be made equal to that inB, when the valve will assume its original position, allowing the air inDto escape.

The force with which the brake is applied depends upon the reduction of pressure in the train pipe. A slight reduction would admit air very slowly fromBtoD, whereas a full escape from the train pipe would open the valve to its utmost. We have not represented the means whereby the valve is rendered sensitive to these changes, for the reason given above.

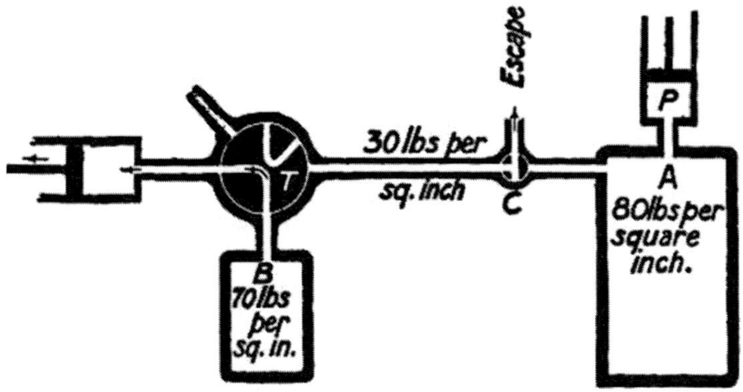

Fig. 89.—Brake "on."

The latest form of triple-valve includes a device which, when air is rapidly discharged from the train pipe, as in an emergency application of the brake, opens a port through which compressed air is also admitted from the train pipe *directly* intoD. It will easily be understood that a double advantage is hereby gained—first, in utilizing a considerable portion of the air in the train pipe to increase the available brake force in cases of emergency; and, secondly, in producing a quick reduction of pressure in the whole length of the pipe, which accelerates the action of the brakes with extraordinary rapidity.

It may be added that this secondary communication is kept open only until the pressure inDis equal to that in the train pipe. Then it is cut off, to prevent a return of air fromBto the pipe.

An interesting detail of the system is the automatic regulation of air-pressure in the main reservoir by the air-pump governor (Fig. 90). The governor is attached to the steam-pipe leading from the locomotive boiler to the air-pump. Steam from the boiler, entering atF, flows through valve 14 and passes byDinto the pump, which is thus brought into operation, and continues to work until the pressure in the main reservoir, acting on the under side of the diaphragm 9, exceeds the tension to which the regulating spring 7 is set. Any excess of pressure forces the diaphragm upwards, lifting valve 11, and allowing compressed air from the main reservoir to flow into the chamberC. The air-pressure forces piston 12 downwards and closes steam-valve 14, thus cutting off the supply of steam to the pump. As soon as the pressure in the reservoir is reduced (by leakage or use) below the normal, spring 7 returns diaphragm 9 to the position shown in Fig. 90, and pin-valve 11 closes. The compressed air previously admitted to the chamberCescapes through the small port *a* to the atmosphere. The steam, acting on the lower surface of valve 14, lifts it and its piston to the position shown, and again flows to the pump, which works until the required air-pressure is again obtained in the reservoir.

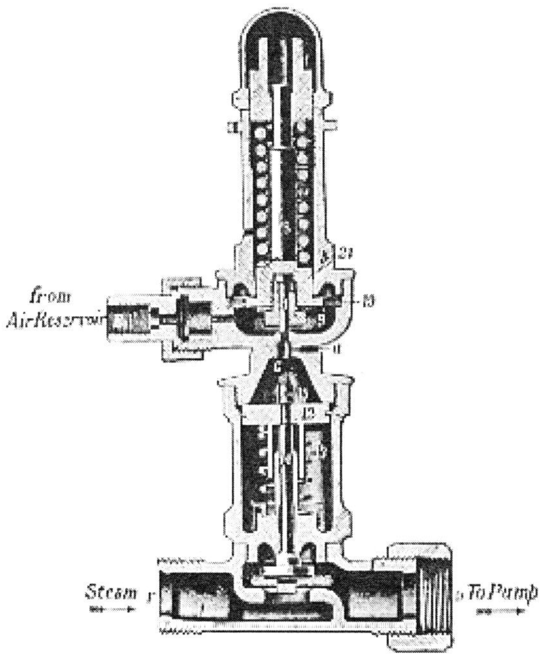

Fig. 90.—Air-pump of Westinghouse brake.

[21] This resembles the upper part of the rudimentary water injector shown in Fig. 15. The reader need only imagine pipeBto be connected with the train pipe. A rush of steam through pipeAcreates a partial vacuum in the coneE, causing air from the train pipe to rush into it and be expelled by the steam blast.

CHAPTER XI.
RAILWAY SIGNALLING.

The block system—Position of signals—Interlocking the signals—Locking gear—Points—Points and signals in combination—Working the block system—Series of signalling operations—Single line signals—The train staff—Train staff and ticket—Electric train staff system—Interlocking—Signalling operations—Power signalling—Pneumatic signalling—Automatic signalling.

UNDER certain conditions—namely, at sharp curves or in darkness—the most powerful brakes might not avail to prevent a train running into the rear of another, if trains were allowed to follow each other closely over the line. It is therefore necessary to introduce an effective system of keeping trains running in the same direction a sufficient distance apart, and this is done by giving visible and easily understood orders to the driver while a train is in motion.

In the early days of the railway it was customary to allow a time interval between the passings of trains, a train not being permitted to leave a station until at least five minutes after the start of a preceding train. This method did not, of course, prevent collisions, as the first train sometimes broke down soon after leaving the station; and in the absence of effective brakes, its successor ran into it. The advent of the electric telegraph, which put stations in rapid communication with one another, proved of the utmost value to the safe working of railways.

THE BLOCK SYSTEM.

Time limits were abolished and distance limits substituted. A line was divided into *blocks*, or lengths, and two trains going in the same direction were never allowed on any one block at the same time.

The signal-posts carrying the movable arms, or semaphores, by means of which the signalman communicates with the engine-driver, are well known to us. They are usually placed on the left-hand side of the line of rails to which they apply, with their arms pointing away from the rails. The side of the arms which faces the direction from which a train approaches has a white stripe painted on a red background, the other side has a black stripe on a white background.

The distant and other signal arms vary slightly in shape (Fig. 91). A distant signal has a forked end and aV-shaped stripe; the home and starting signals are square-ended, with straight stripes. When the arm stands horizontally, the signal is "on," or at "danger"; when dropped, it is "off," and indicates "All right; proceed." At the end nearest the post it carries a spectacle frame glazed with panes of red and green glass. When the arm is at danger, the red pane is opposite a lamp attached to the signal post; when the arm drops, the green pane rises to that position—so that a driver is kept as fully informed at night as during the day, provided the lamp remains alight.

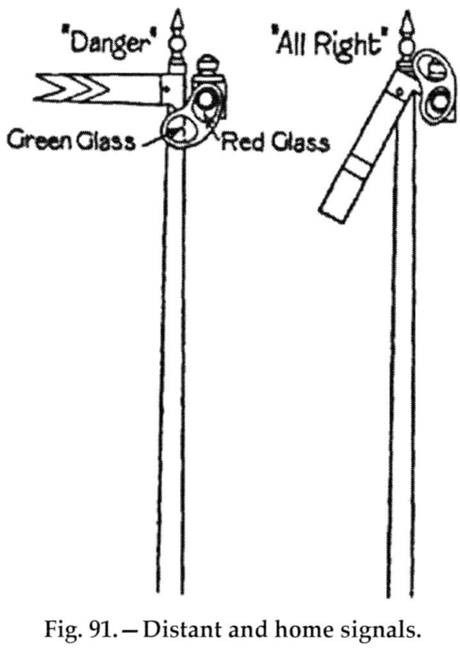

Fig. 91.—Distant and home signals.

POSITION OF SIGNALS.

On double lines each set of rails has its own separate signals, and drivers travelling on the "up" line take no notice of signals meant for the "down" line. Each signal-box usually controls three signals on each set of rails—the distant, the home, and the starting. Their respective positions will be gathered from Fig. 92, which shows a station on a double line. Between the distant and the home an interval is allowed of 800 yards on the level, 1,000 yards on a falling gradient, and 600 yards on a rising gradient. The home stands near the approach end of the station, and the starting at the departure end of the platform. The last is sometimes reinforced by an "advance starting" signal some distance farther on.

It should be noted that the distant is only a *caution* signal, whereas both home and starting are *stop* signals. This means that when the driver sees the distant "on," he does not stop his train, but slackens speed, and prepares to stop at the home signal. He must, however, on no account pass either home or starting if they are at danger. In short, the distant merely warns the driver of what he may expect at the home. To prevent damage if a driver should overrun the home, it has been laid down that no train shall be allowed to pass the starting signal of one box unless the line is clear to a point at least a quarter of a mile beyond the home of the next box. That point is called the *standard clearing point*.

Technically described, a *block* is a length of line between the last stop signal worked from one signal-box and the first stop signal worked from the next signal-box in advance.

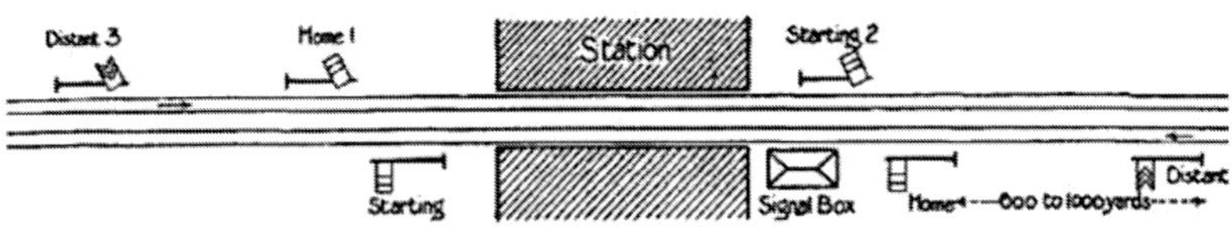

Fig. 92.—Showing position of signals. Those at the top are "off."

INTERLOCKING SIGNALS.

A signalman cannot lower or restore his signals to their normal positions in any order he likes. He is compelled to lower them as follows:—Starting and home; *then* distant. And restore them—distant; *then* starting and home. If a signalman were quite independent, he might, after the passage of a train, restore the home or starting, but forget all about the distant, so that the next train, which he wants to stop, would dash past the distant without warning and have to pull up suddenly when the home came in sight. But by a mechanical arrangement he is prevented from restoring the home or starting until the distant is at danger; and, *vice versâ,* he cannot lower the last until the other two are off. This mechanism is called *locking gear*.

LOOKING GEAR.

There are many different types of locking gear in use. It is impossible to describe them all, or even to give particulars of an elaborate locking-frame of any one type. But if we confine ourselves to the simplest combination of a stud-locking apparatus, such as is used in small boxes on the Great Western

Railway, the reader will get an insight into the general principles of these safety devices, as the same principles underlie them all.

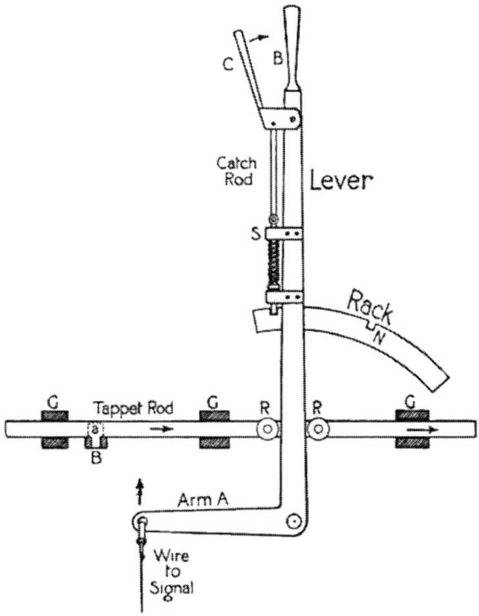

Fig. 93.—A signal lever and its connections. To move the lever,Cis pressed towardsBraising the catch-rod from its nick in the rack, G G G, guides;R R, anti-friction rollers;S, sockets for catch-rod to work in.

The levers in the particular type of locking gear which we are considering have each a tailpiece or "tappet arm" attached to it, which moves backwards and forwards with the lever (Fig. 93). Running at right angles to this tappet, and close to it, either under or above, are the lock bars, or stud bars. Refer now to Fig. 94, which shows the ends of the three tappet arms,D,H, andS, crossed by a bar,B, from which project these studs. The levers are all forward and the signals all "on." If the signalman tried to pull the lever attached toDdown the page, as it were, he would fail to move it on account of the stud *a*, which engages with a notch inD. Before this stud can be got free of the notch the tappetsHandSmust be pulled over, so as to bring their notches in line with studs *b* and *c* (Fig. 95). The signalman can now moveD, since the notch easily pushes the stud *a* to the left (Fig. 96). The signals must be restored to danger. AsHandSare back-locked byD—that is, prevented byDfrom being put back into their normal positions—Dmust be moved first. The interlocking of the three signals described is merely repeated in the interlocking of a large number of signals.

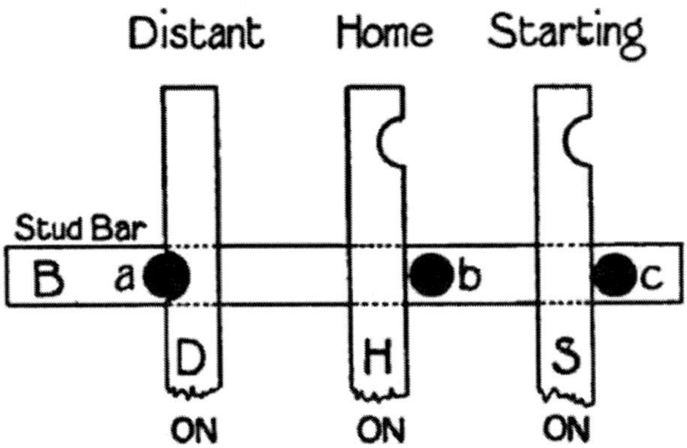

Fig. 94.

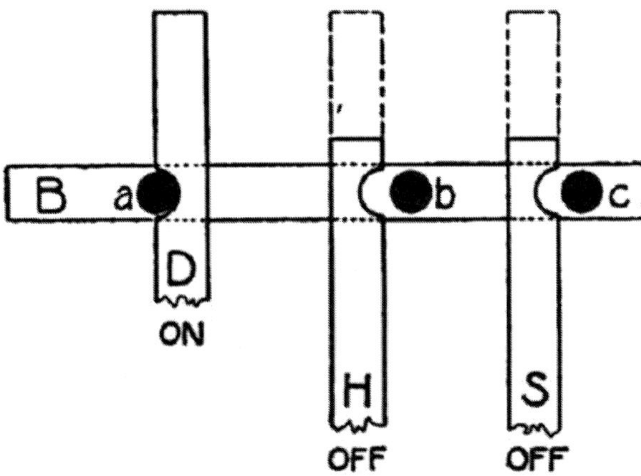

Fig. 95.

On entering a signal-box a visitor will notice that the levers have different colours:— *Green*, signifying distant signals; *red*, signifying home and starting signals; *blue*, signifying facing points; *black*, signifying trailing points; *white*, signifying spare levers. These different colours help the signalman to pick out the right levers easily.

To the front of each lever is attached a small brass tablet bearing certain numbers; one in large figures on the top, then a line, and other numbers in small figures beneath. The large number is that of the lever itself; the others, called *leads*, refer to levers which must be pulled before that particular lever can be released.

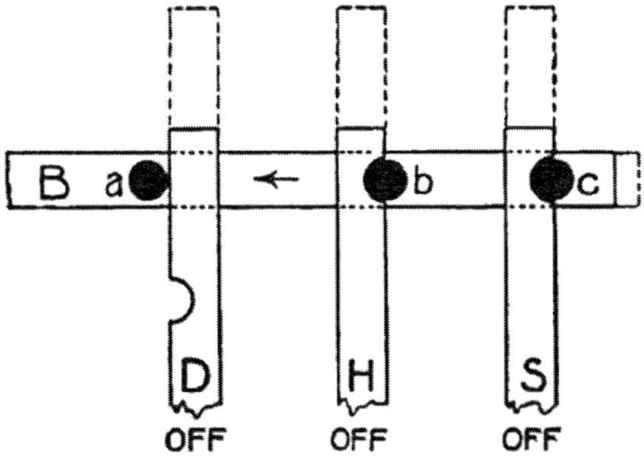

Fig. 96.

Fig. 97.—Model signal equipment in a signalling school. (By permission of the "G.W.R. Magazine").

POINTS.

Mention was made, in connection with the lever, of *points*. Before going further we will glance at the action of these devices for enabling a train to run from one set of rails to another. Figs. 98 and 99 show the points at a simple junction. It will be noticed that the rails of the line to the left of the points are continued as the outer rails of the main and branch lines. The inner rails come to a sharpV-point, and to the left of this are the two short rails which, by means of shifting portions, decide the direction of a train's travel. In Fig. 98 the main line is open; in Fig. 99, the branch. The shifting parts are kept properly spaced by cross bars (or tie-rods),A A.

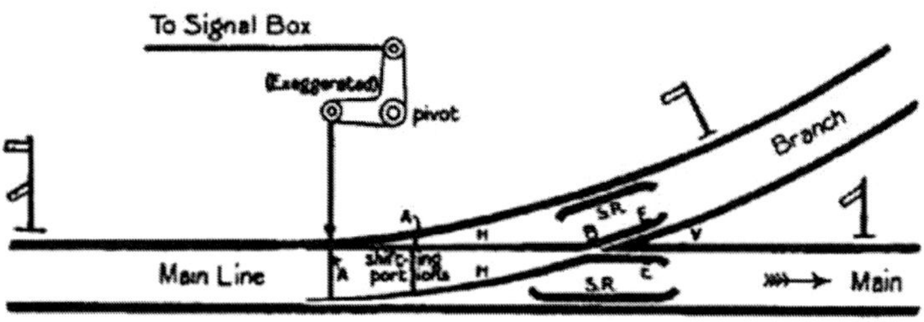

Fig. 98.—Points open to main line.

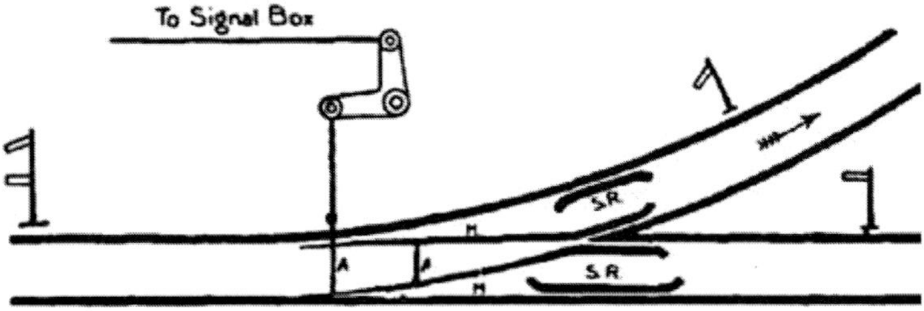

Fig. 99.—Points open to branch line.

It might be thought that the wheels would bump badly when they reach the pointB, where there is a gap. This is prevented, however, by the bent endsE E(Fig. 98), on which the tread of the wheel rests until it has reached some distance along the point ofV. The safety railsS Rkeep the outer wheel up against its rail until theV has been passed.

POINTS AND SIGNALS IN COMBINATION.

Let us suppose that a train is approaching the junction shown in Figs. 98 and 99 from the left. It is not enough that the driver should know that the tracks are clear. He must also be assured that the track, main or branch, as the case may be, along which he has to go, is open; and on the other hand, if he were approaching from the right, he would want to be certain that no train on the other line was converging on his. Danger is avoided and assurance given by interlocking the points and signals. To the left of the junction the home and distant signals are doubled, there being two semaphore arms on each post. These are interlocked with the points in such a manner that the signals referring to either line can be pulled off only when the points are set to open the way to that line. Moreover, before any shifting of points can be made, the signals behind must be put to danger. The convergence of trains is prevented by interlocking, which renders it impossible to have both sets of distant and home signals at "All right" simultaneously.

WORKING OF BLOCK SYSTEM.

We may now pass to the working of the block system of signalling trains from station to station on one line of a double track. Each signal-box (except, of course, those at termini) has electric communication with the next box in both directions. The instruments used vary on different systems, but the principle is the same; so we will concentrate our attention on those most commonly employed on the Great Western Railway. They are:—(1.) Two tapper-bell instruments, connected with similar instruments in the adjacent boxes on both sides. Each of these rings one beat in the corresponding box every time its key is depressed. (2.) Two Spagnoletti disc instruments—one, having two keys, communicating with the box in the rear; and the other, in connection with the forward box, having no keys. Their respective functions are to give signals and receive them. In the centre of the face of each is a square opening, behind which moves a disc carrying two "flags"—"Train on line" in white letters on red ground, and "Line clear" in black letters on a white ground. The keyed instrument has a red and a white key. When the red key is depressed, "Train on line" appears at the opening; also in that of a keyless disc at the adjacent signal-box. A depression of the white key similarly gives "Line clear." A piece of wire with the ends turned over and passed through two eyes slides over the keys, and can be made to hold either down. In addition to these, telephonic and telegraphic instruments are provided to enable the signalmen to converse.

SERIES OF SIGNALLING OPERATIONS.

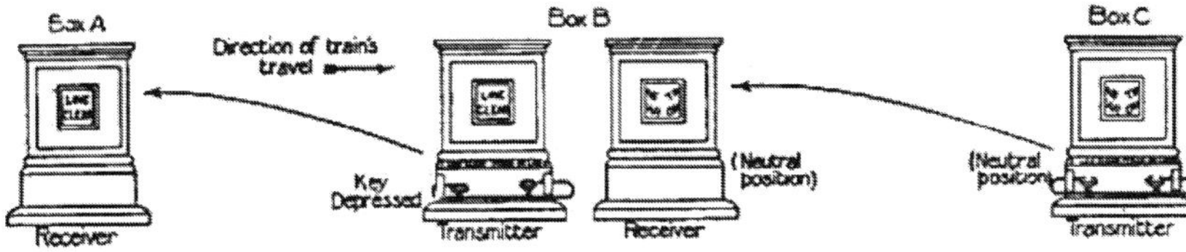

Fig. 100.—The signaling instruments in three adjacent cabins. The featherless arrows show the connection of the instruments.

We may now watch the doings of signalmen in four successive boxes, A, B, C, and D, during the passage of an express train. Signalman A calls signalman B's attention by one beat on the tapper-bell. B answers by repeating it to show that he is attending. A asks, "Is line clear for passenger express?"—four beats on the bell. B, seeing that the line is clear to his clearing point, sends back four beats, and pins down the white key of his instrument. "Line clear" appears on the opening, and also at that of A's keyless disc. A lowers starting signal. Train moves off. A gives two beats on the tapper = "Train entering section." B pins indicator at "Train on line," which also appears on A's instrument. A places signals at danger. B asks C, "Is line clear?" C repeats the bell code, and pins indicator at "Line clear," shown on B's keyless disc also. B lowers all signals. Train passes. B signals to C, "Train entering section." B signals to A, "Train out of section," and releases indicator, which returns to normal position with half of each flag showing at the window. B signals to C, "Train on line," and sets all his signals to danger. C pins indicator to "Train on line." C asks, "Is line clear?" But there is a train at station D, and signalman D therefore gives no reply, which is equivalent to a negative. The driver, on approaching C's distant, sees it at danger, and slows down, stopping at the home. C lowers home, and allows train to proceed to his starting signal. D, when the line is clear to his clearing point, signals "Line clear," and pins indicator at "Line clear." C lowers starting signals, and train proceeds. C signals to D, "Train entering section," and D pins indicator at "Train on line." C signals to B, "Train out of section," sets indicator at normal, and puts signals at danger. And so the process is repeated from station to station. Where, however, sections are short, the signalman is advised one section ahead of the approach of a train by an additional signal signifying, "Fast train approaching." The block indicator reminds the signalman of the whereabouts of the train. Unless his keyless indicator is at normal, he may not ask, "Is line clear?" And until he signals back "Line clear" to the box behind, a train is not allowed to enter his section. In this way a section of line with a full complement of signals is always interposed between any two trains.

THE WORKING OF SINGLE LINES.

We have dealt with the signalling arrangements pertaining to double lines of railway, showing that a system of signals is necessary to prevent a train running into the back of its predecessor. Where trains in both directions pass over a single line, not only has this element of danger to be dealt with, but also the possibility of a train being allowed to enter a section of line from each end *at the same time*. This is effected in several ways, the essence of each being that the engine-driver shall have in his possession *visible* evidence of the permission accorded him by the signalman to enter a section of single line.

A SINGLE TRAIN STAFF.

The simplest form of working is to allocate to the length of line a "train staff"—a piece of wood about 14 inches long, bearing the names of the stations at either end. This is adopted where only one engine is used for working a section, such as a short branch line. In a case like this there is obviously no danger of two trains meeting, and the train staff is merely the authority to the driver to start a journey. No telegraphic communication is necessary with such a system, and signals are placed only at the ends of the line.

TRAIN STAFF AND TICKET.

On long lengths of single line where more than one train has to be considered, the line is divided into blocks in the way already described for double lines, and a staff is assigned to each, the staffs for the various blocks differing from each other in shape and colour. The usual signals are provided at each station, and block telegraph instruments are employed, the only difference being that one disc, of the key pattern, is used for trains in both directions. On such a line it is, of course, possible that two or more trains may require to follow each other without any travelling intermediately in the opposite direction. This would be impossible if the staff passed uniformly to and fro in the block section; but it is arranged by the introduction of a train staff *ticket* used in conjunction with the staff.

No train is permitted to leave a staff station unless the staff for the section of line to be traversed is at the station; and the driver has the strictest possible instructions that he must *see* the staff. If a second train is required to follow, the staff is *shown* to the driver, and a train staff ticket handed him as his authority to proceed. If, however, the next train over the section will enter from the opposite end, the staff is *handed* to the driver.

To render this system as safe as possible, train staff tickets are of the same colour and shape as the staff for the section to which they apply, and are kept in a special box at the stations, the key being attached to the staff and the lock so arranged that the key cannot be withdrawn unless the box has been locked.

ELECTRIC TRAIN STAFF AND TABLET SYSTEMS.

These systems of working are developments of the last mentioned, by which are secured greater safety and ease in working the line. On some sections of single line circumstances often necessitate the running of several trains in one direction without a return train. For such cases the train staff ticket was introduced; but even on the best regulated lines it is not always possible to secure that the staff shall be at the station where it is required at the right time, and cases have arisen where, no train being available at the station where the staff was, it had to be taken to the other station by a man on foot, causing much delay to traffic. The electric train staff and tablet systems overcome this difficulty. Both work on much the same principle, and we will therefore describe the former.

Fig. 101.—An electric train staff holder:S S, staffs in the slot of the instrument. Leaning against the side of the cabin is a staff showing the keyKat the end for unlocking a siding points between two stations. The engine driver cannot remove the staff until the points have been locked again.

At each end of a block section a train staff instrument (Fig. 101) is provided. In the base of these instruments are a number of train staffs, any one of which would be accepted by an engine-driver as permission to travel over the single line. The instruments are electrically connected, the mechanism securing that a staff can be withdrawn only by the co-operation of the signalman at each end of the section; that, when *all* the staffs are in the instruments, a staff may be withdrawn at *either* end; that, when a staff has been withdrawn, another cannot be obtained until the one out has been restored to one or other of the instruments. The safety of such a system is obvious, as also the assistance to the working by having a staff available for a train no matter from which end it is to enter the section.

The mechanism of the instruments is quite simple. A double-poled electro-magnet is energized by the depression of a key by the signalman at the further end of the block into which the train is to run, and by the turning of a handle by the signalman who requires to withdraw a staff. The magnet, being energized, is able to lift a mechanical lock, and permits the withdrawal of a staff. In its passage through the instrument the staff revolves a number of iron discs, which in turn raise or lower a switch controlling the electrical connections. This causes the electric currents actuating the electro-magnet to oppose each other, the magnetism to cease, and the lock to fall back, preventing another staff being withdrawn. It will naturally be asked, "How is the electrical system restored?" We have said that there were a number of staffs in each instrument—in other words, a given number of staffs, usually twenty, is assigned to the section. Assume that there are ten in each instrument, and that the switch in each is in its lower position. Now withdraw a staff, and one instrument has an odd, the other an even, number of staffs, and similarly one switch is raised while the other remains lowered, therefore the electrical circuit is "out of phase"—that is, the currents in the magnets of each staff instrument are opposed to one another, and cannot release the lock. The staff travels through the section and is placed in the instrument at the other end, bringing the number of staffs to eleven—an odd number, and, what is more important, *raising* the switch. Both switches are now raised, consequently the electric currents will support each other, so that a staff may be withdrawn. Briefly, then, when there is an odd number of staffs in one instrument and an even number in the other, as when a staff is in use, the signalmen are unable to obtain a staff, and consequently cannot give authority for a train to enter the section; but when there is either an odd or an even number of staffs in each instrument a staff may be withdrawn at either end on the co-operation of the signalmen.

We may add that, where two instruments are in the same signal-box, one for working to the box in advance, the other to the rear, it is arranged that the staffs pertaining to one section shall not fit the instrument for the other, and must be of different colours. This prevents the driver accidentally accepting a staff belonging to one section as authority to travel over the other.

INTERLOCKING.

The remarks made on the interlocking of points and signals on double lines apply also to the working of single lines, with the addition that not only are the distant, home, and starting signals interlocked with each other, but with the signals and points governing the approach of a train from the opposite direction—in other words, the signals for the approach of a train to a station from one direction cannot be lowered unless those for the approach to the station of a train from the opposite direction are at danger, and the points correctly set.

SIGNALLING OPERATIONS.

In the working of single lines, as of double, the signalman at the station from which a train is to proceed has to obtain the consent of the signalman ahead, the series of questions to be signalled being very similar to those detailed for double lines. There is, however, one notable exception. On long lengths of single line it is necessary to make arrangements for trains to pass each other. This is done by providing loop lines at intervals, a second pair of rails being laid for the accommodation of one train while another in the opposite direction passes it. To secure that more than one train shall not be on a section of single line between two crossing-places it is laid down that, when a signalman at a non-crossing station is asked to allow a train to approach his station, he must not give permission until he has notified the signalman ahead of him, thus securing that he is not asking permission for trains to approach from both directions at the same time. Both for single and double line working a number of rules designed to deal with cases of emergency are laid down, the guiding principle being safety; but we have now dealt with all the conditions of everyday working, and must pass to the consideration of

Fig. 102.—An electric lever-frame in a signalling cabin at Didcot.

"POWER" SIGNALLING.

In a power system of signalling the signalman is provided with some auxiliary means—electricity, compressed air, etc.—of moving the signals or points under his control. It is still necessary to have a locking-frame in the signal-box, with levers interlocked with each other, and connections between the box and the various points and signals. But the frame is much smaller than an ordinary manual frame, and but little force is needed to move the little levers which make or break an electric circuit, or open an air-valve, according to the power-agent used.

ELECTRIC SIGNALLING.

Fig. 102 represents the locking-frame of a cabin at Didcot, England, where an all-electric system has been installed. Wires lead from the cabin to motors situated at the points and signals, which they operate through worm gearing. When a lever is moved it closes a circuit and sets the current flowing through a motor, the direction of the flow (and consequently of the motor's revolution) depending on whether the lever has been moved forward or backward. Indicators arranged under the levers tell the signalman when the desired movements at the points and signals have been completed. If any motion is not carried through, owing to failure of the current or obstruction of the working parts, an electric lock prevents him continuing operations. Thus, suppose he has to open the main line to an express, he is obliged by the mechanical locking-frame to set all the points correctly before the signals can be lowered. He might move all the necessary levers in due order, yet one set of points might remain open, and, were the signals lowered, an accident would result. But this cannot happen, as the electric locks worked by the points in question block the signal levers, and until the failure has been set right, the signals must remain at "danger."

The point motors are connected direct to the points; but between a signal motor and its arm there is an "electric slot," consisting of a powerful electro-magnet which forms a link in the rod work. To lower a signal it is necessary that the motor shall revolve and a control current pass round the magnet to give it the requisite attractive force. If no control current flows, as would happen were any pair of points not in their proper position, the motor can have no effect on the signal arm to lower it, owing to the magnet letting go its grip. Furthermore, if the signal had been already lowered when the control current failed, it would rise to "danger" automatically, as all signals are weighted to assume the danger position by gravity. The signal control currents can be broken by the signalman moving a switch, so that in case of emergency all signals may be thrown simultaneously to danger.

PNEUMATIC SIGNALLING.

In England and the United States compressed air is also used to do the hard labour of the signalman for him. Instead of closing a circuit, the signalman, by moving a lever half-way over, admits air to a pipe running along the track to an air reservoir placed beside the points or signal to which the lever relates. The air opens a valve and puts the reservoir in connection with a piston operating the points or signal-arm, as the case may be. This movement having been performed, another valve in the reservoir is opened, and air passes back through a second pipe to the signal-box, where it opens a third valve controlling a piston which completes the movement of the lever, so showing the signalman that the operation is complete. With compressed air, as with electricity, a mechanical locking-frame is of course used.

AUTOMATIC SIGNALLING.

To reduce expense, and increase the running speed on lines where the sections are short, the train is sometimes made to act as its own signalman. The rails of each section are all bonded together so as to be in metallic contact, and each section is insulated from the two neighbouring sections. At the further end of a section is installed an electric battery, connected to the rails, which lead the current back to a magnet operating a signal stationed some distance back on the preceding section. As long as current flows the signal is held at "All right." When a train enters the section the wheels and axles short-circuit the current, so that it does not reach the signal magnet, and the signal rises to "danger," and stays there until the last pair of wheels has passed out of the section. Should the current fail or a vehicle break loose and remain on the section, the same thing would happen.

The human element can thus be practically eliminated from signalling. To make things absolutely safe, a train should have positive control over a train following, to prevent the driver overrunning the signals. On electric railways this has been effected by means of contacts working in combination with the signals, which either cut the current off from the section preceding that on which a train may be, or raise a trigger to strike an arm on the train following and apply its brakes.

CHAPTER XII.
OPTICS.

Lenses—The image cast by a convex lens—Focus—Relative position of object and lens—Correction of lenses for colour—Spherical aberration—Distortion of image—The human eye—The use of spectacles—The blind spot.

LIGHT is a third form of that energy of which we have already treated two manifestations—heat and electricity. The distinguishing characteristic of ether light-waves is their extreme rapidity of vibration, which has been calculated to range from 700 billion movements per second for violet rays to 400 billion for red rays.

If a beam of white light be passed through a prism it is resolved into the seven visible colours of the spectrum—violet, indigo, blue, green, yellow, orange, and red—in this order. The human eye is most sensitive to the yellow-red rays, a photographic plate to the green-violet rays.

All bodies fall into one of two classes—(1) *Luminous*—that is, those which are a *source* of light, such as the sun, a candle flame, or a red-hot coal; and (2) *non-luminous*, which become visible only by virtue of light which they receive from other bodies and reflect to our eyes.

THE PROPAGATION OF LIGHT.

Light naturally travels in a straight line. It is deflected only when it passes from one transparent medium into another—for example, from air to water—and the mediums are of different densities. We may regard the surface of a visible object as made up of countless points, from each of which a diverging pencil of rays is sent off through the ether.

LENSES.

If a beam of light encounters a transparent glass body with non-parallel sides, the rays are deflected. The direction they take depends on the shape of the body, but it may be laid down as a rule that they are bent toward the thicker part of the glass. The common burning-glass is well known to us.

We hold it up facing the sun to concentrate all the heat rays that fall upon it into one intensely brilliant spot, which speedily ignites any inflammable substance on which it may fall (Fig. 103). We may imagine that one ray passes from the centre of the sun through the centre of the glass. This is undeflected; but all the others are bent towards it, as they pass through the thinner parts of the lens.

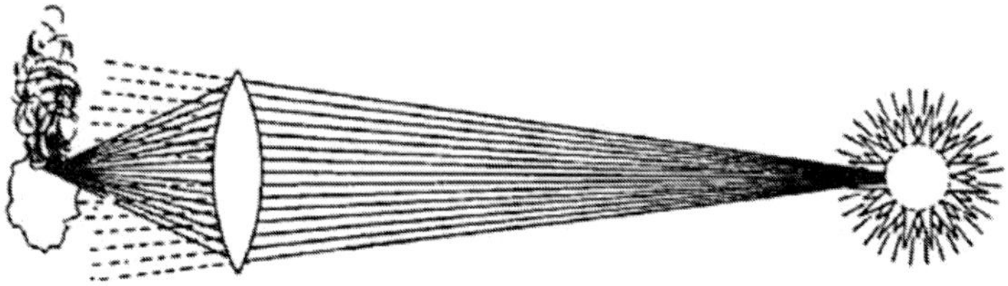

Fig. 103.—Showing how a burning-glass concentrates the heat rays which fall upon it.

It should be noted here that *sunlight*, as we call it, is accompanied by heat. A burning-glass is used to concentrate the *heat* rays, not the *light* rays, which, though they are collected too, have no igniting effect.

In photography we use a lens to concentrate light rays only. Such heat rays as may pass through the lens with them are not wanted, and as they have no practical effect are not taken any notice of. To be of real value, a lens must be quite symmetrical—that is, the curve from the centre to the circumference must be the same in all directions.

There are six forms of simple lenses, as given in Fig. 104. Nos. 1 and 2 have one flat and one spherical surface. Nos. 3, 4, 5, 6 have two spherical surfaces. When a lens is thicker at the middle than at the sides it is called a *convex* lens; when thinner, a *concave* lens. The names of the various shapes are as follows:—No. 1, plano-convex; No. 2, plano-concave; No. 3, double convex; No. 4, double concave; No. 5, meniscus; No. 6, concavo-convex. The thick-centre lenses, as we may term them (Nos. 1, 3, 5), *concentrate* a pencil of rays passing through them; while the thin-centre lenses (Nos. 2, 4, 6) *scatter* the rays (see Fig. 105).

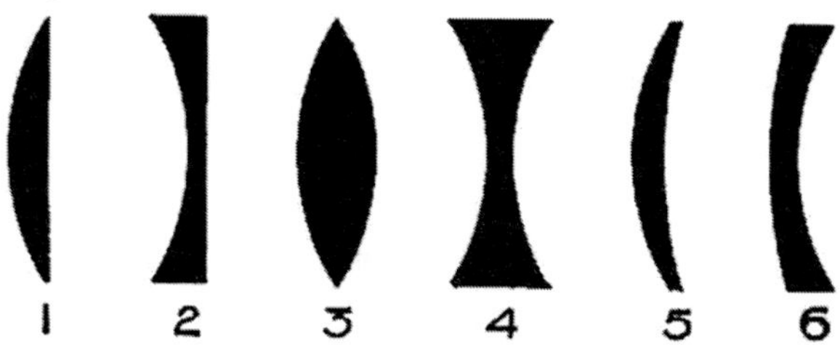

Fig. 104.—Six forms of lenses.

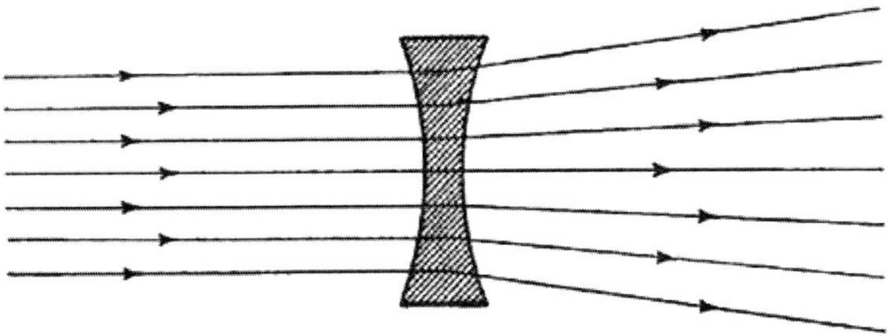

Fig. 105.

Fig. 106.

We said above that light is propagated in straight lines. To prove this is easy. Get a piece of cardboard and prick a hole in it. Set this up some distance away from a candle flame, and hold behind it a piece of tissue paper. You will at once perceive a faint, upside-down image of the flame on the tissue. Why is this? Turn for a moment to Fig. 106, which shows a "pinhole" camera in section. At the rear is a ground-glass screen,B, to catch the image. Suppose thatAis the lowest point of the flame. A pencil of rays diverging from it strikes the front of the camera, which stops them all except the one which passes through the hole and makes a tiny luminous spot onB, *above* the centre of the screen, thoughAis below the axis of the camera. Similarly the tip of the flame (above the axis) would be represented by a dot on the screen below its centre. And so on for all the millions of points of the flame. If we were to enlarge the hole we should get a brighter image, but it would have less sharp outlines, because a number of rays from every point of the candle would reach the screen and be jumbled up with the rays of neighbouring pencils. Now, though a good, sharp photograph may be taken through a pinhole, the time required is so long that photography of this sort has little practical value. What we want is a large hole for the light to enter the camera by, and yet to secure a distinct image. If we place a lens in the hole we can fulfil our wish. Fig. 107 shows a lens in position, gathering up a number of rays from a point,A, and focussing them on a point,B. If the lens has 1,000 times the area of the pinhole, it will pass 1,000 times as many rays, and the image ofAwill be impressed on a sensitized photographic plate 1,000 times more quickly.

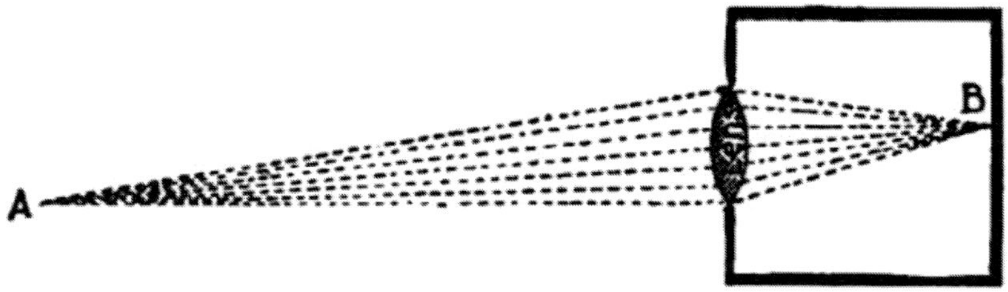

Fig. 107.

THE IMAGE CAST BY A CONVEX LENS.

Fig. 108 shows diagrammatically how a convex lens forms an image. FromAandB, the extremities of the object, a simple ray is considered to pass through the centre of the lens. This is not deflected at all. Two other rays from the same points strike the lens above and below the centre respectively. These are bent inwards and meet the central rays, or come to a focus with them at A1 and B1. In reality a countless number of rays would be transmitted from every point of the object and collected to form the image.

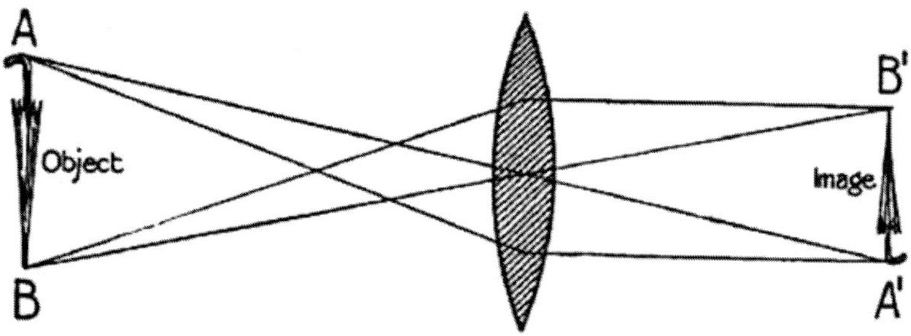

Fig. 108.—Showing how an image is cast by a convex lens.

FOCUS.

We must now take special notice of that word heard so often in photographic talk—"focus." What is meant by the focus or focal length of a lens? Well, it merely signifies the distance between the optical centre of the lens and the plane in which the image is formed.

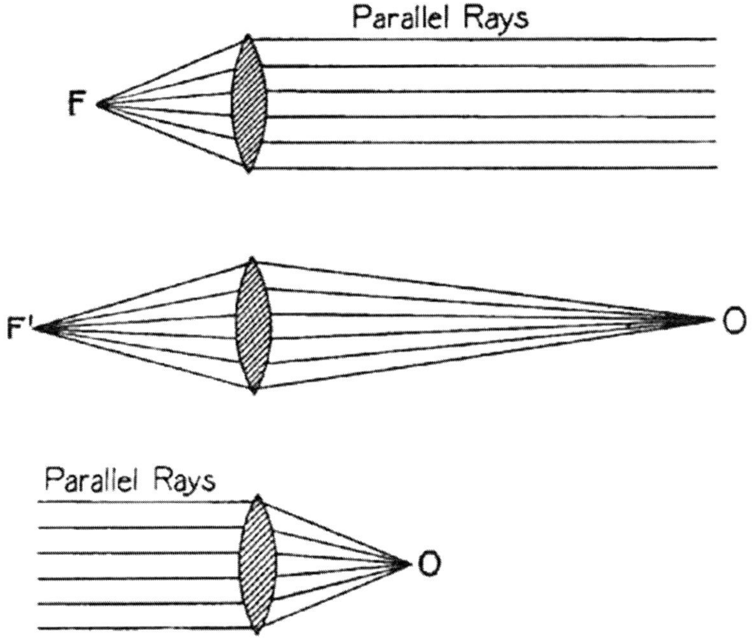

Fig. 109.

We must here digress a moment to draw attention to the three simple diagrams of Fig. 109. The object,O, in each case is assumed to be to the right of the lens. In the topmost diagram the object is so far away from the lens that all rays coming from a single point in it are practically parallel. These converge to a focus atF. If the distance betweenFand the centre of the lens is six inches, we say that the lens has a six-inch focal length. The focal length of a lens is judged by the distance between lens and image when the object is far away. To avoid confusion, this focal length is known as the *principal* focus, and is denoted by the symbol *f*. In the middle diagram the object is quite near the lens, which has to deal with rays striking its nearer surface at an acuter angle than before (reckoning from the centre). As the lens can only deflect their path to a fixed degree, they will not, after passing the lens, come together until they have reached a point,F1, further from the lens thanF. The nearer we approachOto the lens, the further away on the other side is the focal point, until a distance equal to that ofFfrom the lens is reached, when the rays emerge from the glass in a parallel pencil. The rays now come to a focus no longer, and there can be no image. IfObe brought nearer than the focal distance, the rays would *diverge* after passing through the lens.

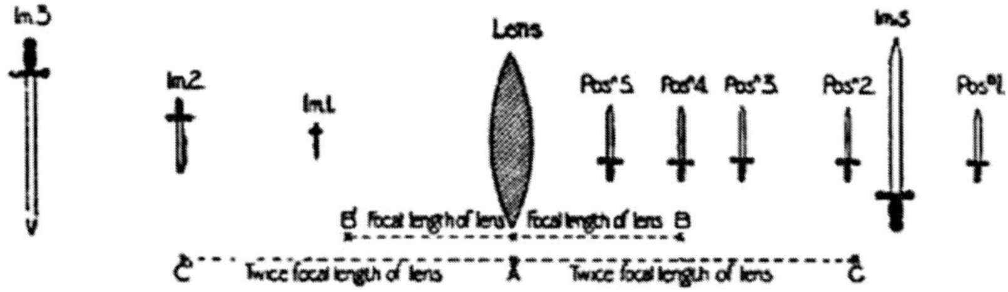

Fig. 110.—Showing how the position of the image alters relatively to the position of the object.

From what has been said above we deduce two main conclusions—(1.) The nearer an object is brought to the lens, the further away from the lens will the image be. (2.) If the object approaches within the principal focal distance of the lens, no image will be cast by the lens. To make this plainer we append a diagram (Fig. 110), which shows five positions of an object and the relative positions of the image (in dotted lines). First, we note that the lineA B, orA B1, denotes the principal focal length of the lens, andA C, orA C1, denotes twice the focal length. We will take the positions in order:—

Position I. Object further away than 2 *f*. Inverted image *smaller* than object, at distance somewhat exceeding *f*.

Position II. Object at distance = 2 *f*. Inverted image at distance = 2 *f*, and of size equal to that of object.

Position III Object nearer than 2 *f*. Inverted image further away than 2 *f*; *larger* than the object.

Position IV. Object at distance = *f*. As rays are parallel after passing the lens *no* image is cast.

Position V. Object at distance less than *f*. No real image—that is, one that can be caught on a focussing screen—is now given by the lens, but a magnified, erect, *virtual* image exists on the same side of the lens as the object.

We shall refer to *virtual* images at greater length presently. It is hoped that any reader who practises photography will now understand why it is necessary to rack his camera out beyond the ordinary focal distance when taking objects at close quarters. From Fig. 110 he may gather one practically useful hint—namely, that to copy a diagram, etc., full size, both it and the plate must be exactly 2 *f* from the optical centre of the lens. And it follows from this that the further he can rack his camera out beyond 2 f the greater will be the possible enlargement of the original.

CORRECTION OF LENSES FOR COLOUR.

We have referred to the separation of the spectrum colours of white light by a prism. Now, a lens is one form of prism, and therefore sorts out the colours. In Fig. 111 we assume that two parallel red rays and two parallel violet rays from a distant object pass through a lens. A lens has most bending effect on violet rays and least on red, and the other colours of the spectrum are intermediately influenced. For the sake of simplicity we have taken the two extremes only. You observe that the pointR, in which the red rays meet, is much further from the lens than isV, the meeting-point of the violet rays. A photographer very seldom has to take a subject in which there are not objects of several different colours, and it is obvious that if he used a simple lens like that in Fig. 111 and got his red objects in good focus, the blue and green portions of his picture would necessarily be more or less out of focus.

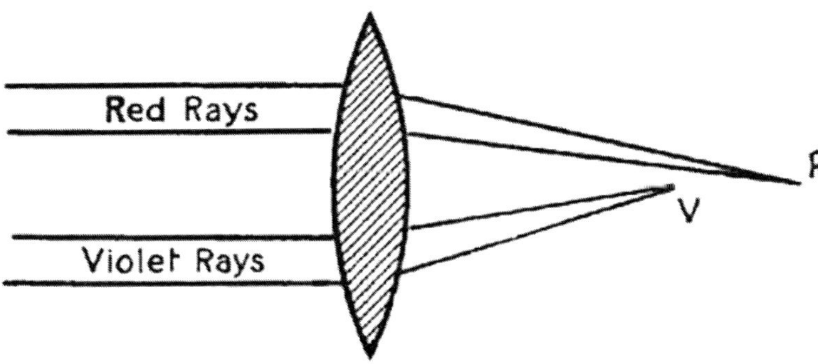

Fig. 111.

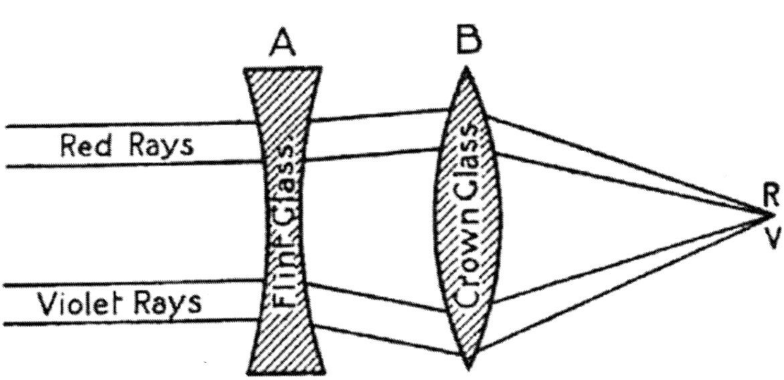

Fig. 112.

This defect can fortunately be corrected by the method shown in Fig. 112. A *compound* lens is needed, made up of a *crown* glass convex element,B, and a concave element,A, of *flint* glass. For the sake of illustration the two parts are shown separated; in practice they would be cemented together, forming one optical body, thicker in the centre than at the edges—a meniscus lens in fact, sinceAis not so concave asBis convex. Now, it was discovered by a Mr. Hall many years ago that if white light passed through two similar prisms, one of flint glass the other of crown glass, the former had the greater effect in separating the spectrum colours—that is, violet rays were bent aside more suddenly compared with the red rays than happened with the crown-glass prism. Look at Fig. 112. The red rays passing through the flint glass are but little deflected, while the violet rays turn suddenly outwards. This is just what is wanted, for it counteracts the unequal inward refraction byB, and both sets of rays come to a focus in the same plane. Such a lens is called *achromatic*, or colourless. If you hold a common reading-glass some distance away from large print you will see that the letters are edged with coloured bands, proving that the lens is not achromatic. A properly corrected photographic lens would not show these pretty edgings. Colour correction is necessary also for lenses used in telescopes and microscopes.

SPHERICAL ABERRATION.

A lens which has been corrected for colour is still imperfect. If rays pass through all parts of it, those which strike it near the edge will be refracted more than those near the centre, and a blurred focus results. This is termed *spherical aberration*. You will be able to understand the reason from Figs. 113 and 114. Two rays,A, are parallel to the axis and enter the lens near the centre (Fig. 113). These meet in one plane. Two other rays,B, strike the lens very obliquely near the edge, and on that account are both turned sharply upwards, coming to a focus in a plane nearer the lens thanA. If this happened in a camera the results would be very bad. EitherAorBwould be out of focus. The trouble is minimized by placing in front of the lens a plate with a central circular opening in it (denoted by the thick, dark line in Fig. 114). The raysBof Fig. 113 are stopped by this plate, which is therefore called a *stop*. But other rays from the same point pass through the hole. These, however, strike the lens much more squarely above the centre, and are not unduly refracted, so that they are brought to a focus in the same plane as raysA.

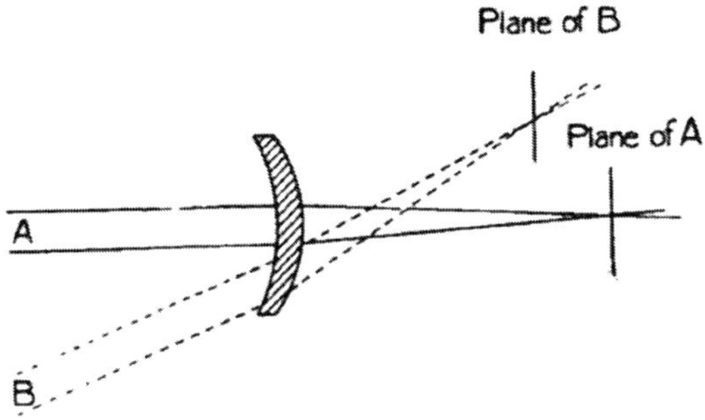

Fig. 113.

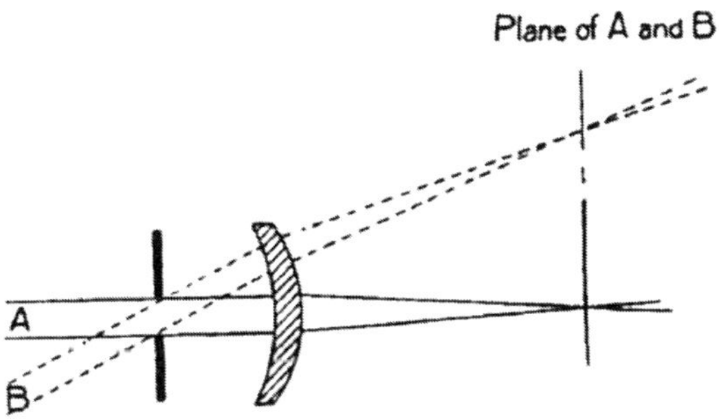

Fig. 114.

DISTORTION OF IMAGE.

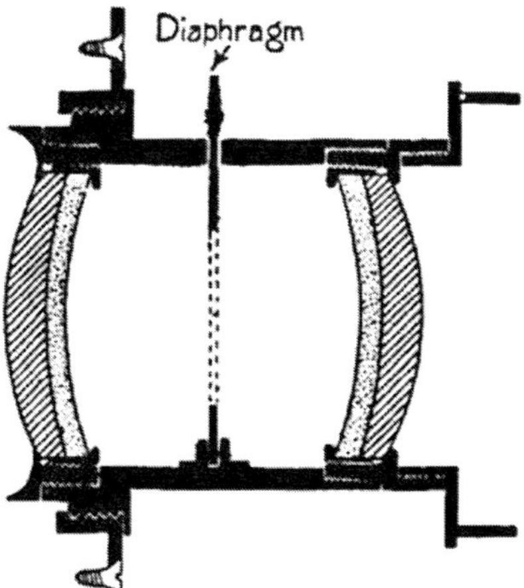

Fig. 115.—Section of a rectilinear lens.

The lens we have been considering is a single meniscus, such as is used in landscape photography, mounted with the convex side turned towards the inside of the camera, and having the stop in front of it. If you possess a lens of this sort, try the following experiment with it. Draw a large square on a sheet of white paper and focus it on the screen. The sides instead of being straight bow outwards: this is called *barrel* distortion. Now turn the lens mount round so that the lens is outwards and the stop inwards. The sides of the square will appear to bow towards the centre: this is *pin-cushion* distortion. For a long time opticians were unable to find a remedy. Then Mr. George S. Cundell suggested that *two* meniscus lenses should be used in combination, one on either side of the stop, as in Fig 115. Each produces distortion, but it is counteracted by the opposite distortion of the other, and a square is represented as a square. Lenses of this kind are called *rectilinear*, or straight-line producing.

We have now reviewed the three chief defects of a lens—chromatic aberration, spherical aberration, and distortion—and have seen how they may be remedied. So we will now pass on to the most perfect of cameras,

THE HUMAN EYE.

The eye (Fig. 116) is nearly spherical in form, and is surrounded outside, except in front, by a hard, horny coat called the *sclerotica* (S). In front is the *cornea* (A), which bulges outwards, and acts as a transparent window to admit light to the lens of the eye (C). Inside the sclerotica, and next to it, comes the *choroid* coat; and inside that again is the *retina*, or curved focussing screen of the eye, which may best be described as a network of fibres ramifying from the optic nerve, which carries sight sensations to the brain. The hollow of the ball is full of a jelly-like substance called the *vitreous humour*; and the cavity between the lens and the cornea is full of water.

We have already seen that, in focussing, the distance between lens and image depends on the distance between object and lens. Now, the retina cannot be pushed nearer to or pulled further away from its lens, like the focussing screen of a camera. How, then, is the eye able to focus sharply objects at distances varying from a foot to many miles?

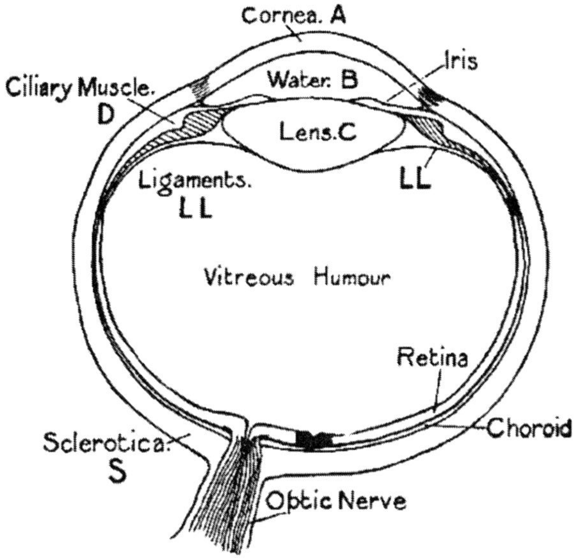

Fig. 116.—Section of the human eye.

As a preliminary to the answer we must observe that the more convex a lens is, the shorter is its focus. We will suppose that we have a box camera with a lens of six-inch focus fixed rigidly in the position necessary for obtaining a sharp image of distant objects. It so happens that we want to take with it a portrait of a person only a few feet from the lens. If it were a bellows camera, we should rack out the back or front. But we cannot do this here. So we place in front of our lens a second convex lens which shortens its principal focus; so that *in effect* the box has been racked out sufficiently.

Nature, however, employs a much more perfect method than this. The eye lens is plastic, like a piece of india-rubber. Its edges are attached to ligaments (L L), which pull outwards and tend to flatten the curve of its surfaces. The normal focus is for distant objects. When we read a book the eye adapts itself to the work. The ligaments relax and the lens decreases in diameter while thickening at the centre, until its curvature is such as to focus all rays from the book sharply on the retina. If we suddenly look through the window at something outside, the ligaments pull on the lens envelope and flatten the curves.

This wonderful lens is achromatic, and free from spherical aberration and distortion of image. Nor must we forget that it is aided by an automatic "stop," the *iris*, the central hole of which is named the *pupil*. We say that a person has black, blue, or gray eyes according to the colour of the iris. Like the lens, the iris adapts itself to all conditions, contracting when the light is strong, and opening when the light is weak, so that as uniform an amount of light as conditions allow may be admitted to the eye. Most modern camera lenses are fitted with adjustable stops which can be made larger or smaller by twisting a ring on the mount, and are named "iris" stops. The image of anything seen is thrown on the retina upside down, and the brain reverses the position again, so that we get a correct impression of things.

THE USE OF SPECTACLES.

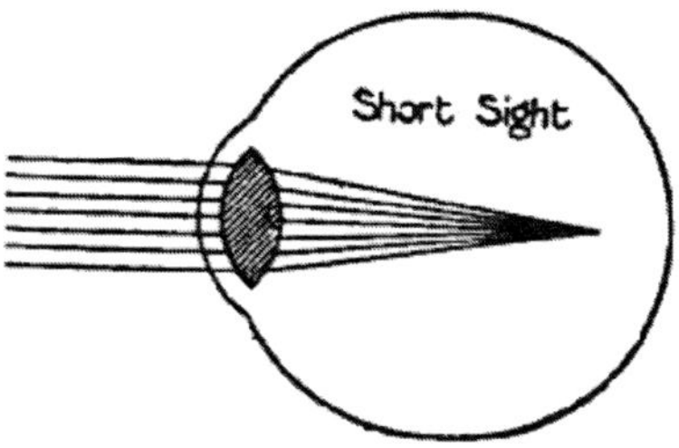

Fig. 117 a.

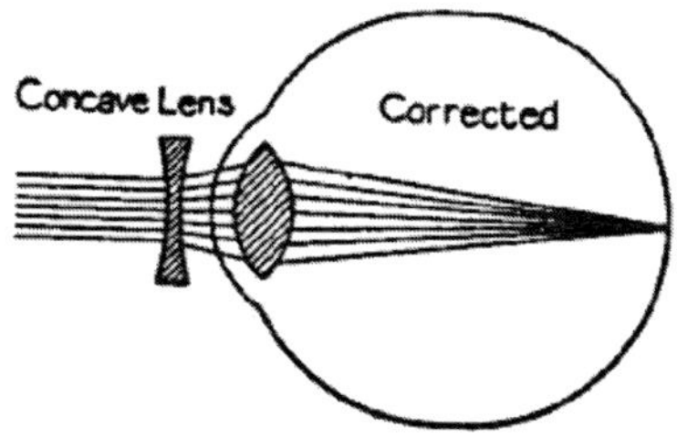

Fig. 117 b.

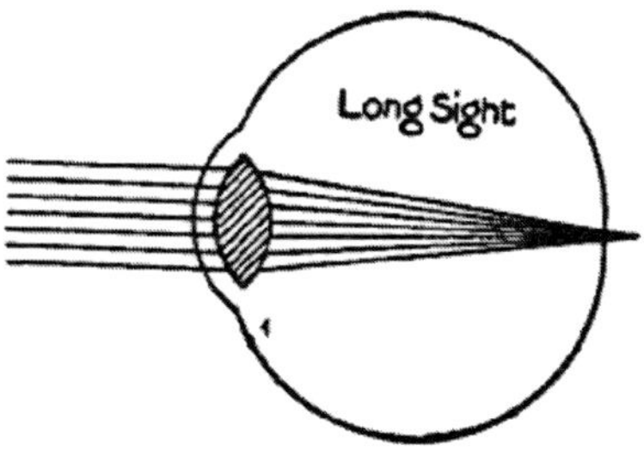

Fig. 118 a.

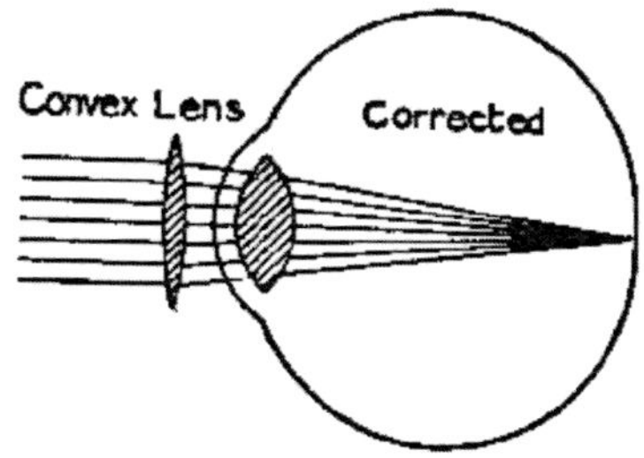

Fig. 118 b.

The reader will now be able to understand without much trouble the function of a pair of spectacles. A great many people of all ages suffer from short-sight. For one reason or another the distance between lens and retina becomes too great for a person to distinguish distant objects clearly. The lens, as shown in Fig 117 *a,* is too convex—has its minimum focus too short—and the rays meet and cross before they reach the retina, causing general confusion of outline. This defect is simply remedied by placing in front of the eye (Fig. 117 *b*) a *concave* lens, to disperse the rays somewhat before they enter the eye, so that they come to a focus on the retina. If a person's sight is thus corrected for distant objects, he can still see near objects quite plainly, as the lens will accommodate its convexity for them. The scientific term for short-sight is *myopia*. Long-sight, or *hypermetropia,* signifies that the eyeball is too short or the lens too flat. Fig. 118 *a* represents the normal condition of a long-sighted eye. When looking at a distant object the eye thickens slightly and brings the focus forward into the retina. But its thickening power in such an eye is very limited, and consequently the rays from a near object focus behind the retina. It is therefore necessary for a long-sighted person to use *convex* spectacles for reading the newspaper. As seen in Fig. 118 *b*, the spectacle lens concentrates the rays before they enter the eye, and so does part of the eye's work for it.

Returning for a moment to the diagram of the eye (Fig. 116), we notice a black patch on the retina near the optic nerve. This is the "yellow spot." Vision is most distinct when the image of the object looked at is formed on this part of the retina. The "blind spot" is that point at which the optic nerve enters the retina, being so called from the fact that it is quite insensitive to light. The finding of the blind spot is an interesting little experiment. On a card make a large and a small spot three inches apart, the one an eighth, the other half an inch in diameter. Bring the card near the face so that an eye is exactly opposite to each spot, and close the eye opposite to the smaller. Now direct the other eye to this spot and you will find, if the card be moved backwards and forwards, that at a certain distance the large spot, though many times larger than its fellow, has completely vanished, because the rays from it enter the open eye obliquely and fall on the "blind spot."

CHAPTER XIII.
THE MICROSCOPE, THE TELESCOPE, AND THE MAGIC-LANTERN.

The simple microscope—Use of the simple microscope in the telescope—The terrestrial telescope—The Galilean telescope—The prismatic telescope—The reflecting telescope—The parabolic mirror—The compound microscope—The magic-lantern—The bioscope—The plane mirror.

IN Fig. 119 is represented an eye looking at a vase, three inches high, situated atA, a foot away. If we were to place another vase,B, six inches high, at a distance of two feet; orC, nine inches high, at three feet; orD, a foot high, at four feet, the image on the retina would in every case be of the same size as that cast byA. We can therefore lay down the rule that *the apparent size of an object depends on the angle that it subtends at the eye.*

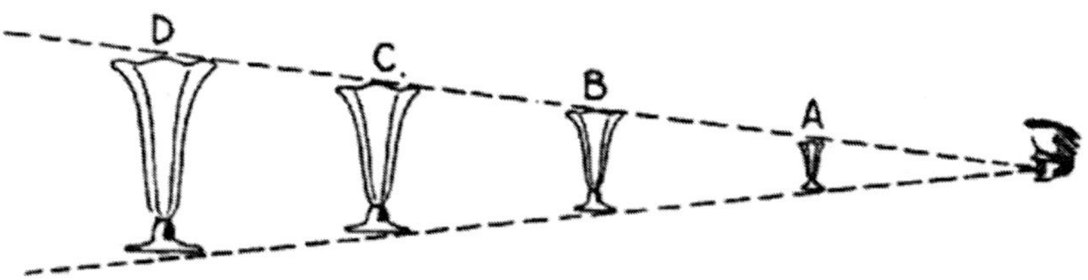

Fig. 119.

To see a thing more plainly, we go nearer to it; and if it be very small, we hold it close to the eye. There is, however, a limit to the nearness to which it can be brought with advantage. The normal eye is unable to adapt its focus to an object less than about ten inches away, termed the "least distance of distinct vision."

THE SIMPLE MICROSCOPE.

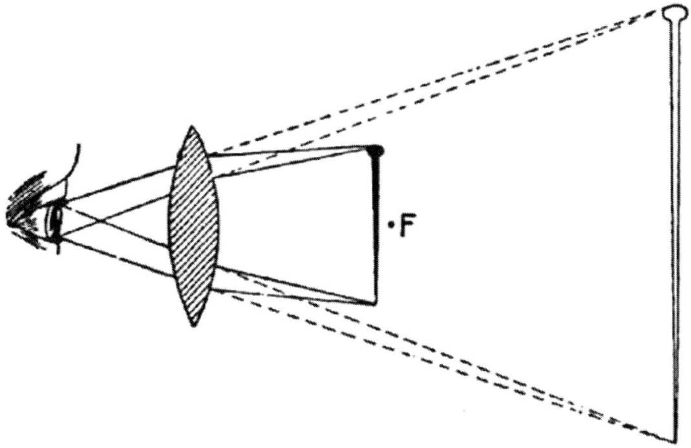

Fig. 120.

A magnifying glass comes in useful when we want to examine an object very closely. The glass is a lens of short focus, held at a distance somewhat less than its principal focal length,F(see Fig. 120), from the object. The rays from the head and tip of the pin which enter the eye are denoted by continuous lines. As they are deflected by the glass the eye gets the *impression* that a much longer pin is situated a considerable distance behind the real object in the plane in which the refracted rays would meet if produced backwards (shown by the dotted lines). The effect of the glass, practically, is to remove it (the object) to beyond the least distance of distinct vision, and at the same time to retain undiminished the angle it subtends at the eye, or, what amounts to the same thing, the actual size of the image formed on the retina.[22] It follows, therefore, that if a lens be of such short focus that it allows us to see an object clearly at a distance of two inches—that is, one-fifth of the least distance of distinct vision—we shall get an image on the retina five times larger in diameter than would be possible without the lens.

The two simple diagrams (Figs. 121 and 122) show why the image to be magnified should be nearer to the lens than the principal focus,F. We have already seen (Fig. 109) that rays coming from a point in the principal focal plane emerge as a parallel pencil. These the eye can bring to a focus, because it normally has a curvature for focussing parallel rays. But, owing to the power of "accommodation," it can also focus *diverging* rays (Fig. 121), the eye lens thickening the necessary amount, and we therefore put our magnifying glass a bit nearer thanFto get full advantage of proximity. If we had the object *outside* the principal focus, as in Fig. 122, the rays from it would converge, and these could not be gathered to a sharp point by the eye lens, as it cannot *flatten* more than is required for focussing parallel rays.

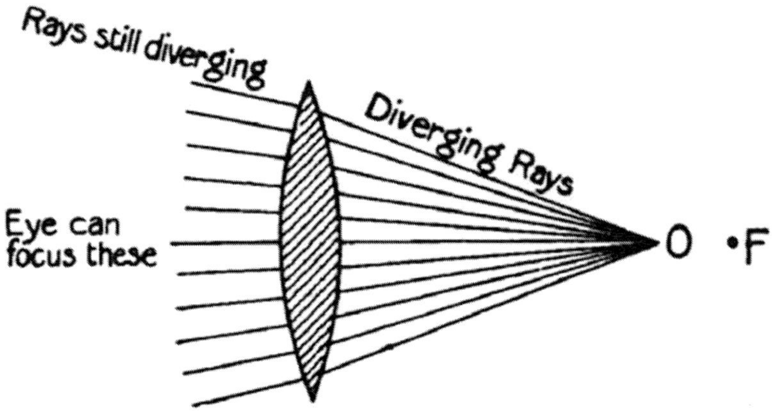

Fig. 121.

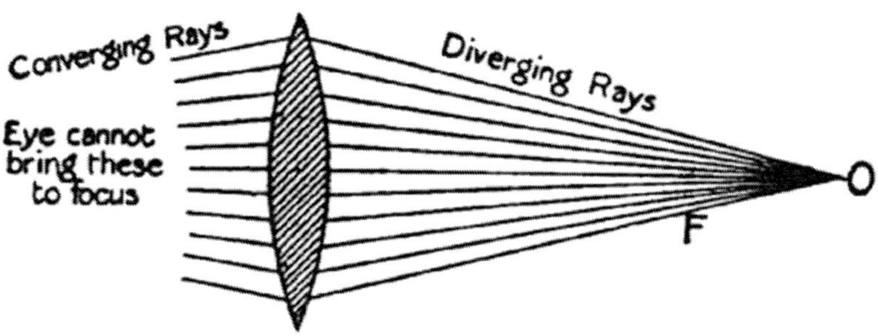

Fig. 122.

USE OF THE SIMPLE MICROSCOPE IN THE TELESCOPE.

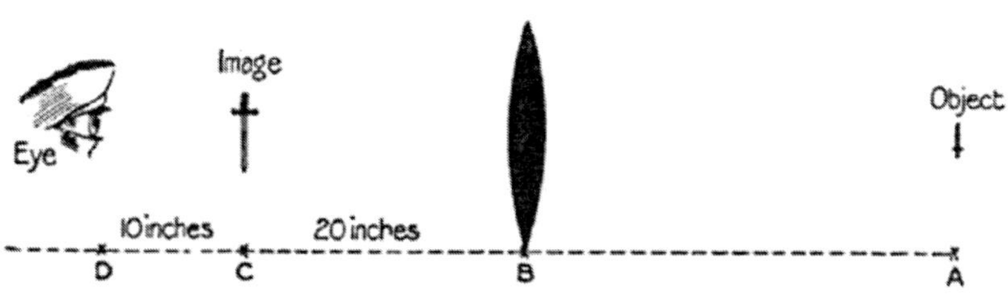

Fig. 123.

Let us now turn to Fig. 123. AtAis a distant object, say, a hundred yards away.Bis a double convex lens, which has a focal length of twenty inches. We may suppose that it is a lens in a camera. An inverted image of the object is cast by the lens atC. If the eye were placed atC, it would distinguish nothing. But if withdrawn toD, the least distance of distinct vision,[23] behindC, the image is seen clearly. That the image really is atCis proved by letting down the focussing screen, which at once catches it. Now, as the focus of the lens is twice *d*, the image will be twice as large as the object would appear if viewed directly without the lens. We may put this into a very simple formula:—

$$\textit{Magnification} = \frac{\text{focal length of lens}}{d}$$

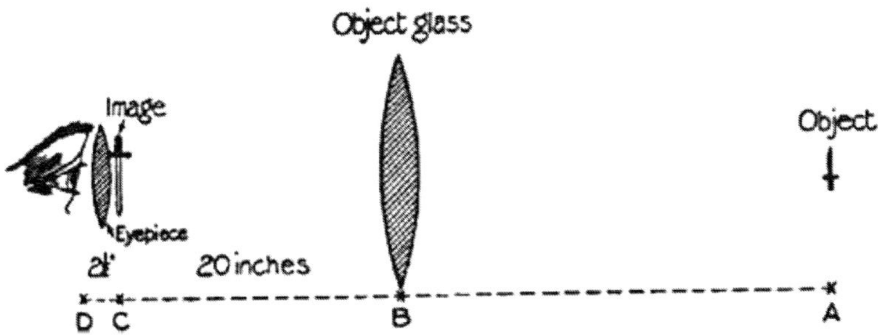

Fig. 124.

In Fig. 124 we have interposed between the eye and the object a small magnifying glass of 2½-inch focus, so that the eye can now clearly see the image when one-quarter *d* away from it.Balready magnifies the image twice; the eye-piece again magnifies it four times; so that the total magnification is 2 × 4 = 8 times. This result is arrived at quickly by dividing the focus ofB(which corresponds to the object-glass of a telescope) by the focus of the eye-piece, thus:—

$$\frac{20}{2\frac{1}{2}} = 8$$

The ordinary astronomical telescope has a very long focus object-glass at one end of the tube, and a very short focus eye-piece at the other. To see an object clearly one merely has to push in or pull out the eye-piece until its focus exactly corresponds with that of the object-glass.

THE TERRESTRIAL TELESCOPE.

An astronomical telescope inverts images. This inversion is inconvenient for other purposes. So the terrestrial telescope (such as is commonly used by sailors) has an eye-piece compounded of four convex lenses which erect as well as magnify the image. Fig. 125 shows the simplest form of compound erecting eye-piece.

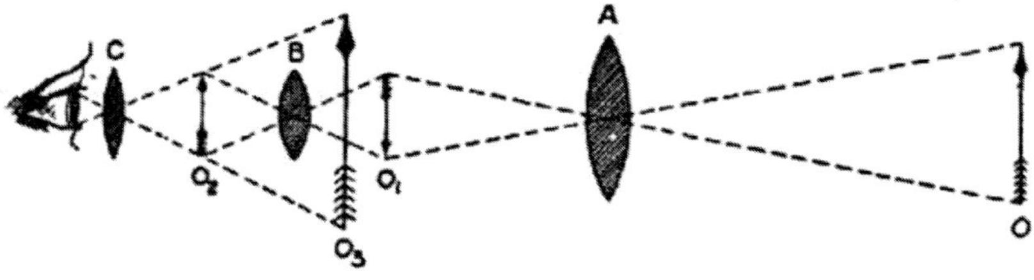

Fig. 125.

THE GALILEAN TELESCOPE.

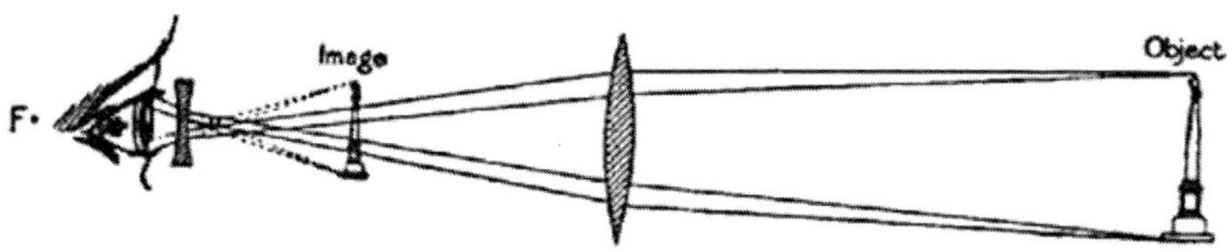

Fig. 126.

A third form of telescope is that invented by the great Italian astronomer, Galileo,[24] in 1609. Its principle is shown in Fig. 126. The rays transmitted by the object-glass are caught, *before* coming to a focus, on a concave lens which separates them so that they appear to meet in the paths of convergence denoted by the dotted lines. The image is erect. Opera-glasses are constructed on the Galilean principle.

THE PRISMATIC TELESCOPE.

In order to be able to use a long-focus object-glass without a long focussing-tube, a system of glass reflecting prisms is sometimes employed, as in Fig. 127. A ray passing through the object-glass is reflected from one posterior surface of prismAon to the other posterior surface, and by it out through the front on to a second prism arranged at right angles to it, which passes the ray on to the compound eye-piece. The distance between object-glass and eye-piece is thus practically trebled. The best-known prismatic telescopes are the Zeiss field-glasses.

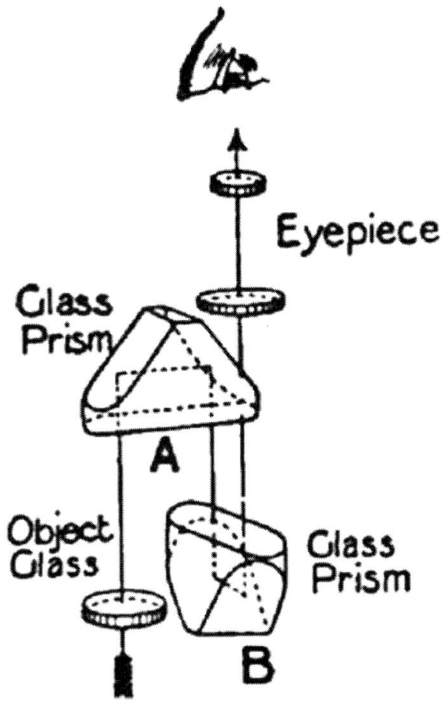

Fig. 127.

THE REFLECTING TELESCOPE.

We must not omit reference to the *reflecting* telescope, so largely used by astronomers. The front end of the telescope is open, there being no object-glass. Rays from the object fall on a parabolic mirror situated in the rear end of the tube. This reflects them forwards to a focus. In the Newtonian reflector a plane mirror or prism is situated in the axis of the tube, at the focus, to reflect the rays through an eye-piece projecting through the side of the tube. Herschel's form of reflector has the mirror set at an angle to the axis, so that the rays are reflected direct into an eye-piece pointing through the side of the tube towards the mirror.

THE PARABOLIC MIRROR.

This mirror (Fig. 128) is of such a shape that all rays parallel to the axis are reflected to a common point. In the marine searchlight a powerful arc lamp is arranged with the arc at the focus of a parabolic reflector, which sends all reflected light forward in a pencil of parallel rays. The most powerful searchlight in existence gives a light equal to that of 350 million candles.

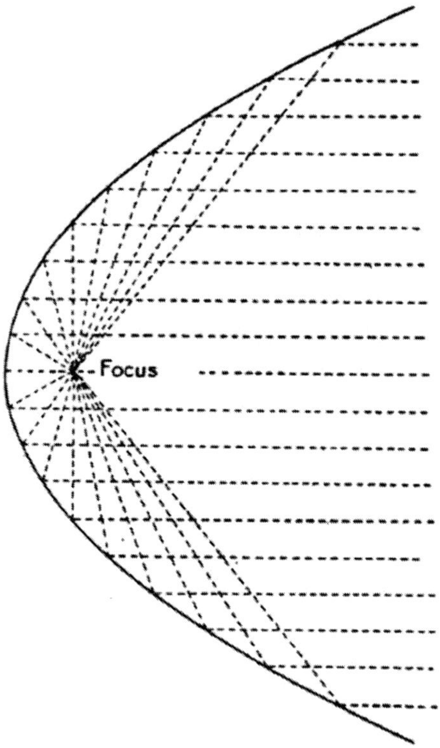

Fig. 128.—A parabolic reflector.

THE COMPOUND MICROSCOPE.

We have already observed (Fig. 110) that the nearer an object approaches a lens the further off behind it is the real image formed, until the object has reached the focal distance, when no image at all is cast, as it is an infinite distance behind the lens. We will assume that a certain lens has a focus of six inches. We place a lighted candle four feet in front of it, and find that a *sharp* diminished image is cast on a ground-glass screen held seven inches behind it. If we now exchange the positions of the candle and the screen, we shall get an enlarged image of the candle. This is a simple demonstration of the law of *conjugate foci*—namely, that the distance between the lens and an object on one side and that between the lens and the corresponding image on the other bear a definite relation to each other; and an object placed at either focus will cast an image at the other. Whether the image is larger or smaller than the object depends on which focus it occupies. In the case of the object-glass of a telescope the image was at what we may call the *short* focus.

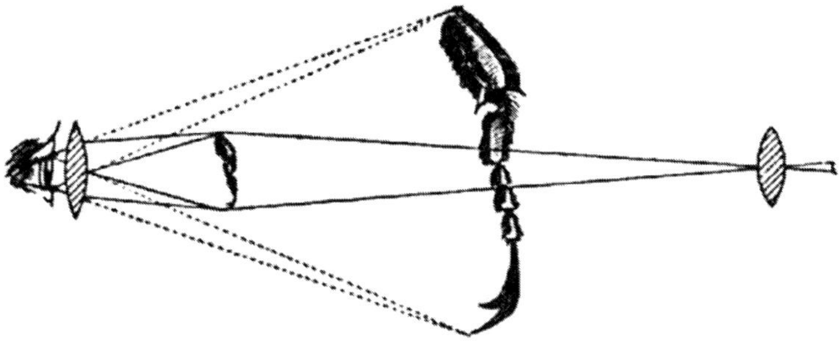

Fig. 129.—Diagram to explain the compound microscope.

Now, a compound microscope is practically a telescope with the object at the *long* focus, very close to a short-focus lens. A greatly enlarged image is thrown (see Fig. 129) at the conjugate focus, and this is caught and still further magnified by the eye-piece. We may add that the object-glass, or *objective*, of a microscope is usually compounded of several lenses, as is also the eye-piece.

THE MAGIC-LANTERN.

The most essential features of a magic-lantern are:—(1) The *source of light*; (2) the *condenser* for concentrating the light rays on to the slide; (3) the *lens* for projecting a magnified image on to a screen.

Fig. 130 shows these diagrammatically. The *illuminant* is most commonly an oil-lamp, or an acetylene gas jet, or a cylinder of lime heated to intense luminosity by an oxy-hydrogen flame. The natural combustion of hydrogen is attended by a great heat, and when the supply of oxygen is artificially increased the temperature of the flame rises enormously. The nozzle of an oxy-hydrogen jet has an interior pipe connected with the cylinder holding one gas, and an exterior, and somewhat larger, pipe leading from that containing the other, the two being arranged concentrically at the nozzle. By means of valves the proportions of the gases can be regulated to give the best results.

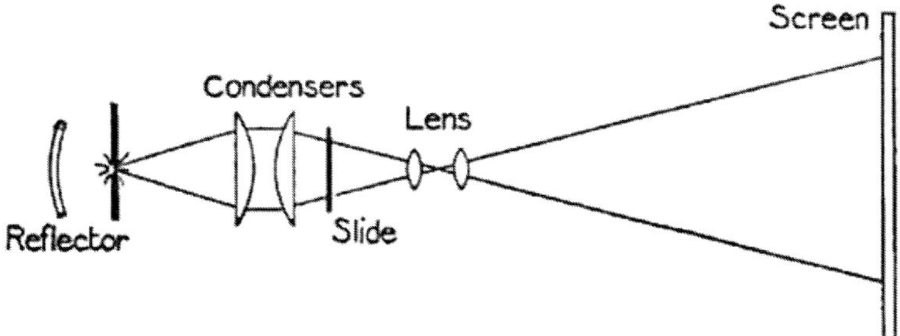

Fig. 130.—Sketch of the elements of a magic-lantern.

The *condenser* is set somewhat further from the illuminant than the principal focal length of the lenses, so that the rays falling on them are bent inwards, or to the slide.

The *objective*, or object lens, stands in front of the slide. Its position is adjustable by means of a rack and a draw-tube. The nearer it is brought to the slide the further away is the conjugate focus and consequently the image. The exhibitor first sets up his screen and lantern, and then finds the conjugate foci of slide and image by racking the lens in or out.

If a very short focus objective be used, subjects of microscopic proportions can be projected on the screen enormously magnified. During the siege of Paris in 1870–71 the Parisians established a balloon and pigeon post to carry letters which had been copied in a minute size by photography. These copies could be enclosed in a quill and attached to a pigeon's wing. On receipt, the copies were placed in a special lantern and thrown as large writing on the screen. Micro-photography has since then made great strides, and is now widely used for scientific purposes, one of the most important being the study of the crystalline formations of metals under different conditions.

THE BIOSCOPE.

"Living pictures" are the most recent improvement in magic-lantern entertainments. The negatives from which the lantern films are printed are made by passing a ribbon of sensitized celluloid through a special form of camera, which feeds the ribbon past the lens in a series of jerks, an exposure being made automatically by a revolving shutter during each rest. The positive film is placed in a lantern, and the intermittent movement is repeated; but now the source of illumination is behind the film, and light passes outwards through the shutter to the screen. In the Urban bioscope the film travels at the rate of fifteen miles an hour, upwards of one hundred exposures being made every second.

The impression of continuous movement arises from the fact that the eye cannot get rid of a visual impression in less than one-tenth of a second. So that if a series of impressions follow one another more rapidly than the eye can rid itself of them the impressions will overlap, and give one of *motion*, if the position of some of the objects, or parts of the objects, varies slightly in each succeeding picture.[25]

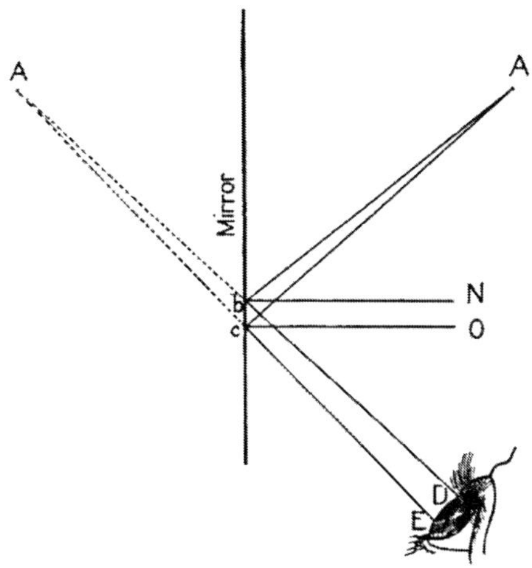

Fig. 131.

This chapter may conclude with a glance at the common looking-glass. Why do we see a reflection in it? The answer is given graphically by Fig. 131. Two rays,A *b*,A *c*, from a pointAstrike the mirrorMat the points *b* and *c*. Lines *b*N, *c*O, drawn from these points perpendicular to the mirror are called their *normals*. The anglesA *b*N, A *c*Oare the *angles of incidence* of raysA *b*,A *c*. The paths which the rays take after reflection must make angles with *b*Nand *c*Orespectively equal toA *b*N,A *c*O. These are the *angles of reflection*. If the eye is so situated that the rays enter it as in our illustration, an image of the pointAis seen at the pointA1, in which the linesD *b*,E *c* meet when produced backwards.

Fig. 132.

When the vertical mirror is replaced by a horizontal reflecting surface, such as a pond (Fig. 132), the same thing happens. The point at which the ray from the reflection of the spire's tip to the eye appears to pass through the surface of the water must be so situated that if a line were drawn perpendicular to it from the surface the angles made by lines drawn from the real spire tip and from the observer's eye to the base of the perpendicular would be equal.

[22] Glazebrook, "Light,".

[23] Glazebrook, "Light,".

[24] Galileo was severely censured and imprisoned for daring to maintain that the earth moved round the sun, and revolved on its axis.

[25] For a full account of Animated Pictures the reader might advantageously consult "The Romance of Modern Invention," foll.

CHAPTER XIV.
SOUND AND MUSICAL INSTRUMENTS.

Nature of sound—The ear—Musical instruments—The vibration of strings—The sounding-board and the frame of a piano—The strings—The striking mechanism—The quality of a note.

SOUND differs from light, heat, and electricity in that it can be propagated through matter only. Sound-waves are matter-waves, not ether-waves. This can be proved by placing an electric bell under the bell-glass of an air-pump and exhausting all the air. Ether still remains inside the glass, but if the bell be set in motion no sound is audible. Admit air, and the clang of the gong is heard quite plainly.

Sound resembles light and heat, however, thus far, that it can be concentrated by means of suitable lenses and curved surfaces. An *echo* is a proof of its *reflection* from a surface.

Before dealing with the various appliances used for producing sound-waves of a definite character, let us examine that wonderful natural apparatus

THE EAR,

through which we receive those sensations which we call sound.

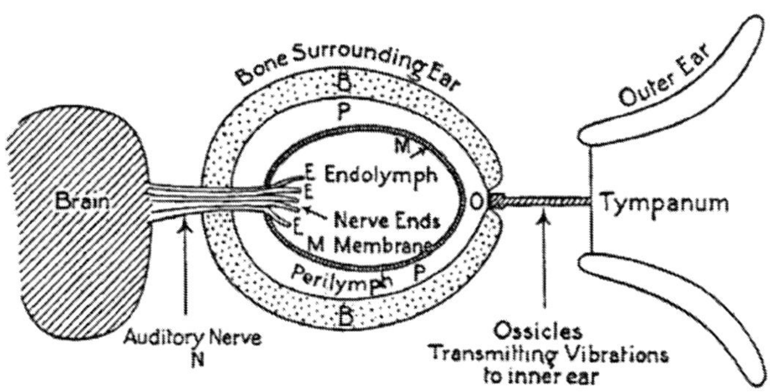

Fig. 133.—Diagrammatic sketch of the parts of the ear.

Fig. 133 is a purely diagrammatic section of the ear, showing the various parts distorted and out of proportion. Beginning at the left, we have the *outer ear*, the lobe, to gather in the sound-waves on to the membrane of the tympanum, or drum, to which is attached the first of a series of *ossicles*, or small bones. The last of these presses against an opening in the *inner ear*, a cavity surrounded by the bones of the head. Inside the inner ear is a watery fluid,P, called *perilymph* ("surrounding water"), immersed in which is a membranic envelope,M, containing *endolymph* ("inside water"), also full of fluid. Into this fluid projectE E E, the terminations of the *auditory nerve*, leading to the brain.

When sound-waves strike the tympanum, they cause it to move inwards and outwards in a series of rapid movements. The ossicles operated by the tympanum press on the little openingO, covered by a membrane, and every time they push it in they slightly squeeze the perilymph, which in turn compresses the endolymph, which affects the nerve-ends, and telegraphs a sensation of sound to the brain.

In Fig. 134 we have a more developed sketch, giving in fuller detail, though still not in their actual proportions, the components of the ear. The ossiclesM,I, andSare respectively the *malleus* (hammer), *incus* (anvil), and *stapes* (stirrup). Each is attached by ligaments to the walls of the middle ear. The tympanum moves the malleus, the malleus the incus, and the incus the stapes, the last pressing into the openingOof Fig. 133, which is scientifically known as the *fenestra ovalis*, or oval window. As liquids are practically incompressible, nature has made allowance for the squeezing in of the oval window membrane, by providing a second opening, the round window, also covered with a membrane. When the stapes pushes the oval membrane in, the round membrane bulges out, its elasticity sufficing to put a certain pressure on the perilymph (indicated by the dotted portion of the inner ear).

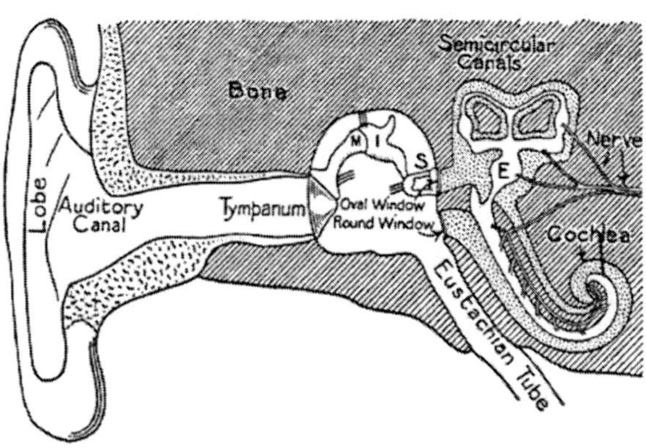

Fig. 134.—Diagrammatic section of the ear, showing the various parts.

The inner ear consists of two main parts, the *cochlea*—so called from its resemblance in shape to a snail's shell—and the *semicircular canals*. Each portion has its perilymph and endolymph, and contains a number of the nerve-ends, which are, however, most numerous in the cochlea. We do not know for certain what the functions of the canals and the cochlea are; but it is probable that the former enables us to distinguish between the *intensity* or loudness of sounds and the direction from which they come, while the latter enables us to determine the *pitch* of a note. In the cochlea are about 2,800 tiny nerve-ends, called the *rods of Corti*. The normal ear has such a range as to give about 33 rods to the semitone. The great scientist Helmholtz has advanced the theory that these little rods are like tiny tuning-forks, each responding to a note of a certain pitch; so that when a string of a piano is sounded and the air vibrations are transmitted to the inner ear, they affect only one of these rods and the part of the brain which it serves, and we have the impression of one particular note. It has been proved by experiment that a very sensitive ear can distinguish between sounds varying in pitch by only1⁄64th of a semitone, or but half the range of any one Corti fibre. This difficulty Helmholtz gets over by suggesting that in such an ear two adjacent fibres are affected, but one more than the other.

A person who has a "good ear" for music is presumably one whose Corti rods are very perfect. Unlucky people like the gentleman who could only recognize one tune, and that because people took off their hats when it commenced, are physically deficient. Their Corti rods cannot be properly developed.

What applies to one single note applies also to the elements of a musical chord. A dozen notes may sound simultaneously, but the ear is able to assimilate each and blend it with its fellows; yet it requires a very sensitive and well-trained ear to pick out any one part of a harmony and concentrate the brain's attention on that part.

The ear has a much larger range than the eye. "While the former ranges over eleven octaves, but little more than a single octave is possible to the latter. The quickest vibrations which strike the eye, as light, have only about twice the rapidity of the slowest; whereas the quickest vibrations which strike the ear, as a musical sound, have more than two thousand times the rapidity of the slowest."[26] To come to actual figures, the ordinary ear is sensitive to vibrations ranging from 16 to 38,000 per second. The bottom and top notes of a piano make respectively about 40 and 4,000 vibrations a second. Of course, some ears, like some eyes, cannot comprehend the whole scale. The squeak of bats and the chirrup of crickets are inaudible to some people; and dogs are able to hear sounds far too shrill to affect the human auditory apparatus.

Not the least interesting part of this wonderful organ is the tympanic membrane, which is provided with muscles for altering its tension automatically. If we are "straining our ears" to catch a shrill sound, we tighten the membrane; while if we are "getting ready" for a deep, loud report like that of a gun, we allow the drum to slacken.

The *Eustachian tube* (Fig. 134) communicates with the mouth. Its function is probably to keep the air-pressure equal on both sides of the drum. When one catches cold the tube is apt to become blocked by mucus, causing unequal pressure and consequent partial deafness.

Before leaving this subject, it will be well to remind our more youthful readers that the ear is delicately as well as wonderfully made, and must be treated with respect. Sudden shouting into the ear, or a playful blow, may have most serious effects, by bursting the tympanum or injuring the arrangement of the tiny bones putting it in communication with the inner ear.

MUSICAL INSTRUMENTS.

These are contrivances for producing sonorous shocks following each other rapidly at regular intervals. Musical sounds are distinguished from mere noises by their regularity. If we shake a number of nails in a tin box, we get only a series of superimposed and chaotic sensations. On the other hand, if we strike a tuning-fork, the air is agitated a certain number of times a second, with a pleasant result which we call a note.

We will begin our excursion into the region of musical instruments with an examination of that very familiar piece of furniture,

THE PIANOFORTE,

which means literally the "soft-strong." By many children the piano is regarded as a great nuisance, the swallower-up of time which could be much more agreeably occupied, and is accordingly shown much less respect than is given to a phonograph or a musical-box. Yet the modern piano is a very clever piece of work, admirably adapted for the production of sweet melody—if properly handled. The two forms of piano now generally used are the *upright*, with vertical sound-board and wires, and the *grand*, with horizontal sound-board.[27]

THE VIBRATION OF STRINGS.

As the pianoforte is a stringed instrument, some attention should be given to the subject of the vibration of strings. A string in a state of tension emits a note when plucked and allowed to vibrate freely. The *pitch* of the note depends on several conditions:—(1) The diameter of the string; (2) the tension of the string; (3) the length of the string; (4) the substance of the string.

Taking them in order:—(1.) The number of vibrations per second is inversely proportional to the diameter of the string: thus, a string one-quarter of an inch in diameter would vibrate only half as often in a given time as a string one-eighth of an inch in diameter. (2.) The length remaining the same, the number of vibrations is directly proportional to the *square root* of the *tension*: thus, a string strained by a 16-lb. weight would vibrate four times as fast as it would if strained by a 1-lb. weight. (3.) The number of vibrations is inversely proportional to the *length* of the string: thus, a one-foot string would vibrate twice as fast as a two-foot string, strained to the same tension, and of equal diameter and weight. (4.) Other things being equal, the rate of vibration is inversely proportional to the square root of the *density* of the substance: so that a steel wire would vibrate more rapidly than a platinum wire of equal diameter, length, and tension. These facts are important to remember as the underlying principles of stringed instruments.

Now, if you hang a wire from a cord, and hang a heavy weight from the wire, the wire will be in a state of high tension, and yield a distinct note if struck. But the volume of sound will be very small, much too small for a practical instrument. The surface of the string itself is so limited that it sets up but feeble motions in the surrounding air. Now hang the wire from a large board and strike it again. The volume of sound has greatly increased, because the string has transmitted its vibrations to the large surface of the board.

To get the full sound-value of the vibrations of a string, we evidently ought to so mount the string that it may influence a large sounding surface. In a violin this is effected by straining the strings over a "bridge" resting on a hollow box made of perfectly elastic wood. Draw the bow across a string. The loud sound heard proceeds not from the string only, but also from the whole surface of the box.

THE SOUNDING-BOARD AND FRAME OF A PIANO.

A piano has its strings strained across a *frame* of wood or steel, from a row of hooks in the top of the frame to a row of tapering square-ended pins in the bottom, the wires passing over sharp edges near both ends. The tuner is able, on turning a pin, to tension its strings till it gives any desired note. Readers may be interested to learn that the average tension of a string is 275 lbs., so that the total strain on the frame of a grand piano is anything between 20 and 30 *tons*.

To the back of the frame is attached the *sounding-board*, made of spruce fir (the familiar Christmas tree). This is obtained from Central and Eastern Europe, where it is carefully selected and prepared, as it is essential that the timber should be sawn in such a way that the grain of the wood runs in the proper direction.

THE STRINGS.

These are made of extremely strong steel wire of the best quality. If you examine the wires of your piano, you will see that they vary in thickness, the thinnest being at the treble end of the frame. It is found impracticable to use wires of the same gauge and the same tension throughout. The makers therefore use highly-tensioned thick wires for the bass, and finer, shorter wires for the treble, taking advantage of the three factors—weight, tension, and length—which we have noticed above. The wires for the deepest notes are wrapped round with fine copper wire to add to their weight without increasing their diameter at the tuning-pins. There are about 600 yards (roughly one-third of a mile) of wire in a grand piano.

THE STRIKING MECHANISM.

We now pass to the apparatus for putting the strings in a state of vibration. The grand piano mechanism shown in Fig. 135 may be taken as typical of the latest improvements. The essentials of an effective mechanism are:—(1) That the blow delivered shall be sharp and certain; (2) that the string shall be immediately "damped," or have its vibration checked if required, so as not to interfere with the succeeding notes of other strings; (3) that the hammer shall be able to repeat the blows in quick succession. The *hammer* has a head of mahogany covered with felt, the thickness of which tapers gradually and regularly from an inch and a quarter at the bass end to three-sixteenths of an inch at the extreme treble notes. The entire eighty-five hammers for the piano are covered all together in one piece, and then they are cut apart from each other. The consistency of the covering is very important. If too hard, it yields a harsh note, and must be reduced to the right degree by pricking with a needle. In the diagram the felt is indicated by the dotted part.

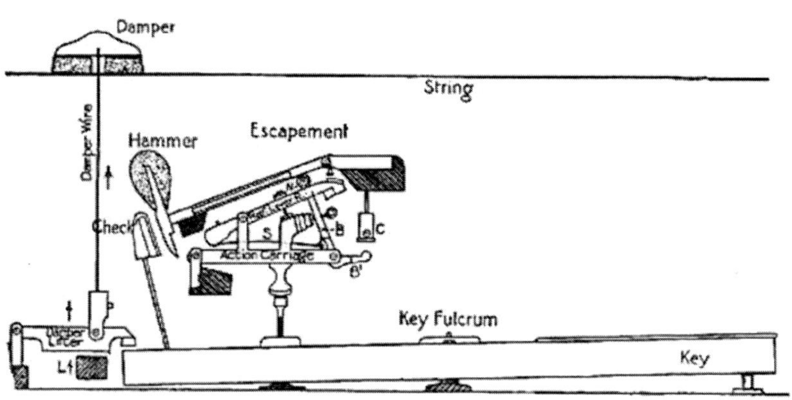

Fig. 135.—The striking mechanism of a "grand" piano.

The *action carriage* which operates the hammer is somewhat complicated. When the key is depressed, the left end rises, and pushes up the whole carriage, which is pivoted at one end. The hammer shank is raised by the jackBpressing upon a knob,N, called the *notch,* attached to the under side of the shank. When the jack has risen to a certain point, its arm,B1, catches against the buttonCand jerks it from under the notch at the very moment when the hammer strikes, so that it may not be blocked against the string. As it rebounds, the hammer is caught on the *repetition lever*R, which lifts it to allow of perfect repetition.

The *check* catches the tail of the hammer head during its descent when the key is raised, and prevents it coming back violently on the carriage and rest. The tail is curved so as to wedge against the check without jamming in any way. The moment the carriage begins to rise, the rear end of the key lifts a lever connected with the *damper* by a vertical wire, and raises the damper of the string. If the key is held down, the vibrations continue for a long time after the blow; but if released at once, the damper stifles them as the hammer regains its seat. A bar,L, passing along under all the *damper lifters,* is raised by depressing the loud pedal. The *soft pedal* slides the whole keyboard along such a distance that the hammers strike two only out of the three strings allotted to all except the bass notes, which have only one string apiece, or two, according to their depth or length. In some pianos the soft pedal presses a special damper against the strings; and a third kind of device moves the hammers nearer the strings so that they deliver a lighter blow. These two methods of damping are confined to upright pianos.

A high-class piano is the result of very careful workmanship. The mechanism of each note must be accurately regulated by its tiny screws to a minute fraction of an inch. It must be ensured that every hammer strikes its blow at exactly the right place on the string, since on this depends the musical value of the note. The adjustment of the dampers requires equal care, and the whole work calls for a sensitive ear combined with skilled mechanical knowledge, so that the instrument may have a light touch, strength, and certainty of action throughout the whole keyboard.

THE QUALITY OF A NOTE.

If two strings, alike in all respects and equally tensioned, are plucked, both will give the same note, but both will not necessarily have the same quality of tone. The quality, or *timbre,* as musicians call it, is influenced by the presence of *overtones,* or *harmonics,* in combination with the *fundamental,* or deepest, tone of the string. The fact is, that while a vibrating string vibrates as a whole, it also vibrates in parts. There are, as it were, small waves superimposed on the big fundamental waves. Points of least motion, called *nodes,* form on the string, dividing it into two, three, four, five, etc., parts, which may be further divided by subsidiary nodes. The string, considered as halved by one node, gives the first overtone, or octave of the fundamental. It may also vibrate as three parts, and give the second overtone, or twelfth of the fundamental;[28] and as four parts, and give the third overtone, the double octave.

Now, if a string be struck at a point corresponding to a node, the overtones which require that point for a node will be killed, on account of the excessive motion imparted to the string at that spot. Thus to hit it at the middle kills the octave, the double octave, etc.; while to hit it at a point one-third of the length from one end stifles the twelfth and all its sub-multiples.

A fundamental note robbed of all its harmonics is hard to obtain, which is not a matter for regret, as it is a most uninteresting sound. To get a rich tone we must keep as many useful harmonics as possible, and therefore a piano hammer is so placed as to strike the string at a point which does not interfere with the best harmonics, but kills those which are objectionable. Pianoforte makers have discovered by experiment that the most pleasing tone is excited when the point against which the hammer strikes is one-seventh to one-ninth of the length of the wire from one end.

The nature of the material which does the actual striking is also of importance. The harder the substance, and the sharper the blow, the more prominent do the harmonics become; so that the worker has to regulate carefully both the duration of the blow and the hardness of the hammer covering.

[26] Tyndall, "On Sound,".

[27] A Broadwood "grand" is made up of 10,700 separate pieces, and in its manufacture forty separate trades are concerned.

[28] Twelve notes higher up the scale.

CHAPTER XV.
WIND INSTRUMENTS.

Longitudinal vibration—Columns of air—Resonance of columns of air—Length and tone—The open pipe—The overtones of an open pipe—Where overtones are used—The arrangement of the pipes and pedals—Separate sound-boards—Varieties of stops—Tuning pipes and reeds—The bellows—Electric and pneumatic actions—The largest organ in the world—Human reeds.

LONGITUDINAL VIBRATION.

IN stringed instruments we are concerned only with the transverse vibrations of a string—that is, its movements in a direction at right angles to the axis of the string. A string can also vibrate longitudinally—that is, in the direction of its axis—as may be proved by drawing a piece of resined leather along a violin string. In this case the harmonics "step up" at the same rate as when the movements were transverse.

Let us substitute for a wire a stout bar of metal fixed at one end only. The longitudinal vibrations of this rod contain overtones of a different ratio. The first harmonic is not an octave, but a twelfth. While a tensioned string is divided by nodes into two, three, four, five, six, etc., parts, a rod fixed at one end only is capable of producing only those harmonics which correspond to division into three, five, seven, nine, etc., parts. Therefore a free-end rod and a wire of the same fundamental note would not have the same *timbre*, or quality, owing to the difference in the harmonics.

COLUMNS OF AIR.

In wind instruments we employ, instead of rods or wires, columns of air as the vibrating medium. The note of the column depends on its length. In the "penny whistle," flute, clarionet, and piccolo the length of the column is altered by closing or opening apertures in the substance encircling the column.

RESONANCE OF COLUMNS OF AIR.

Why does a tube closed at one end, such as the shank of a key, emit a note when we blow across the open end? The act of blowing drives a thin sheet of air against the edge of the tube and causes it to vibrate. The vibrations are confused, some "pulses" occurring more frequently than others. If we blew against the edge of a knife or a piece of wood, we should hear nothing but a hiss. But when, as in the case which we are considering, there is a partly-enclosed column of air close to the pulses, this selects those pulses which correspond to its natural period of vibration, and augments them to a sustained and very audible musical sound.

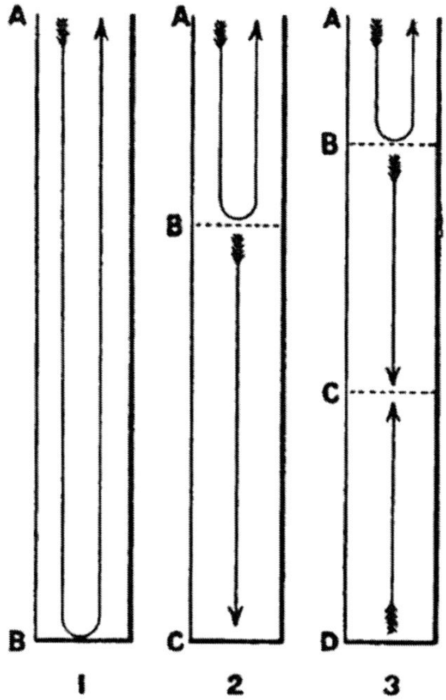

Fig 136.—Showing how the harmonics of a "stopped" pipe are formed.

In Fig. 136, *1* is a pipe, closed at the bottom and open at the top. A tuning-fork of the same note as the pipe is struck and held over it so that the prongs vibrate upwards and downwards. At the commencement of an outward movement of the prongs the air in front of them is *compressed*. This impulse, imparted to the air in the pipe, runs down the column, strikes the bottom, and returns. Just as it reaches the top the prong is beginning to move inwards, causing a *rarefaction* of the air behind it. This effect also travels down and back up the column of air in the pipe, reaching the prong just as it arrives at the furthest point of the inward motion. The process is repeated, and the column of air in the pipe, striking on the surrounding atmosphere at regular intervals, greatly increases the volume of sound. We must observe that if the tuning-fork were of too high or too low a note for the column of air to move in perfect sympathy with it, this increase of sound would not result. Now, when we blow across the end, we present, as it were, a number of vibrating tuning-forks to the pipe, which picks out those air-pulses with which it sympathizes.

LENGTH AND TONE.

The rate of vibration is found to be inversely proportional to the length of the pipe. Thus, the vibrations of a two-foot pipe are twice as rapid as those of a four-foot pipe, and the note emitted by the former is an octave higher than that of the latter. A one-foot pipe gives a note an octave higher still. We are here speaking of the *fundamental* tones of the pipes. With them, as in the case of strings, are associated the *overtones*, or harmonics, which can be brought into prominence by increasing the pressure of the blast at the top of the pipe. Blow very hard on your key, and the note suddenly changes to one much shriller. It is the twelfth of the fundamental, of which it has completely got the upper hand.

We must now put on our thinking-caps and try to understand how this comes about. First, let us note that the vibration of a body (in this case a column of air) means a motion from a point of rest to a point of rest, or from node to node. In the air-column in Fig. 136, *1*, there is only one point of rest for an impulse—namely, at the bottom of the pipe. So that to pass from node to node the impulse must pass up the pipe and down again. The distance from node to node in a vibrating body is called a *ventral segment*. Remember this term. Therefore the pipe represents a semi-ventral segment when the fundamental note is sounding.

When the first overtone is sounded the column divides itself into two vibrating parts. Where will the node between them be? We might naturally say, "Half-way up." But this cannot be so; for if the node were so situated, an impulse going down the pipe would only have to travel to the bottom to find another node, while an impulse going up would have to travel to the top and back again—that is, go twice as far. So the node forms itself *one-third* of the distance down the pipe. FromBtoA(Fig. 136, *2*) and back is now equal to fromBtoC. When the second overtone is blown (Fig. 136, *3*) a third node forms. The pipe is now divided into *five* semi-ventral segments. And with each succeeding overtone another node and ventral segment are added.

The law of vibration of a column of air is that the number of vibrations is directly proportional to the number of semi-ventral segments into which the column of air inside the pipe is divided.[29] If the fundamental tone gives 100 vibrations per second, the first overtone in a closed pipe must give 300, and the second 500 vibrations.

THE OPEN PIPE.

A pipe open at both ends is capable of emitting a note. But we shall find, if we experiment, that the note of a stopped pipe is an octave lower than that of an open pipe of equal length. This is explained by Fig. 137, *1*. The air-column in the pipe (of the same length as that in Fig. 136) divides itself, when an end is blown across, into two equal portions at the nodeB, the natural point to obtain equilibrium. A pulse will pass fromAorA1 toBand back again in half the time required to pass fromAtoBand back in Fig. 136, *1*; therefore the note is an octave higher.

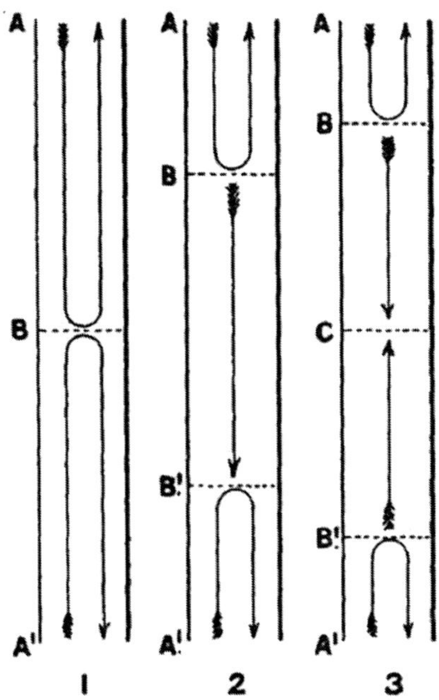

Fig. 137.—Showing how harmonics of an open pipe are formed,B,B1, andCare "nodes." The arrows indicate the distance travelled by a sound impulse from a node to a node.

THE OVERTONES OF AN OPEN PIPE.

The first overtone results when nodes form as in Fig. 137, *2*, at points one-quarter of the length of the pipe from the ends, giving one complete ventral segment and two semi-ventral segments. The vibrations now are twice as rapid as before. The second overtone requires three nodes, as in Fig. 137, *3*. The rate has now trebled. So that, while the overtones of a closed pipe rise in the ratio 1, 3, 5, 7, etc., those of an open pipe rise in the proportion 1, 2, 3, 4, etc.

WHERE OVERTONES ARE USED.

In the flute, piccolo, and clarionet, as well as in the horn class of instrument, the overtones are as important as the fundamental notes. By artificially altering the length of the column of air, the fundamental notes are also altered, while the harmonics of each fundamental are produced at will by varying the blowing pressure; so that a continuous chromatic, or semitonal, scale is possible throughout the compass of the instrument.

THE ORGAN.

From the theory of acoustics[30] we pass to the practical application, and concentrate our attention upon the grandest of all wind instruments, the pipe organ. This mechanism has a separate pipe for every note, properly proportioned. A section of an ordinary wooden pipe is given in Fig. 138. Wind rushes up through the foot of the pipe into a little chamber, closed by a block of wood or a plate except for a narrow slit, which directs it against the sharp lipA, and causes a fluttering, the proper pulse of which is converted by the air-column above into a musical sound.

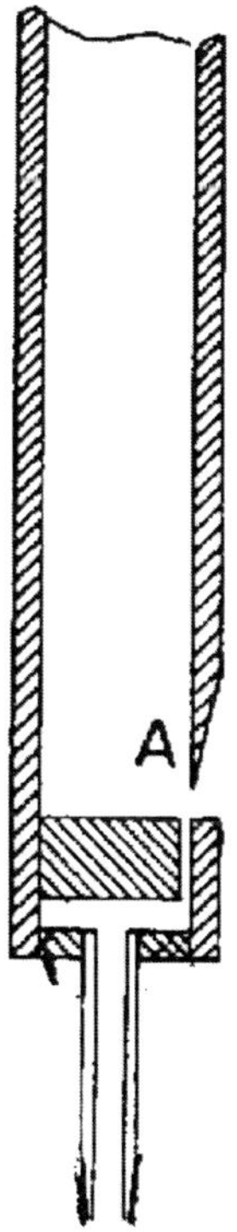

Fig. 138.—Section of an ordinary wooden "flue" pipe.

In even the smallest organs more than one pipe is actuated by one key on the keyboard, for not only do pipes of different shapes give different qualities of tone, but it is found desirable to have ranks of pipes with their bottom note of different pitches. The length of an open pipe is measured from the edge of the lip to the top of the pipe; of a stopped pipe, from the lip to the top and back again. When we speak of a 16 or 8 foot rank, or stop, we mean one of which the lowest note in the rank is that produced by a 16 or 8 foot open pipe, or their stopped equivalents (8 or 4 foot). In a big organ we find 32, 16, 8, 4, and 2 foot stops, and some of these repeated a number of times in pipes of different shape and construction.

THE ARRANGEMENT OF THE PIPES.

We will now study briefly the mechanism of a very simple single-keyboard organ, with five ranks of pipes, or stops.

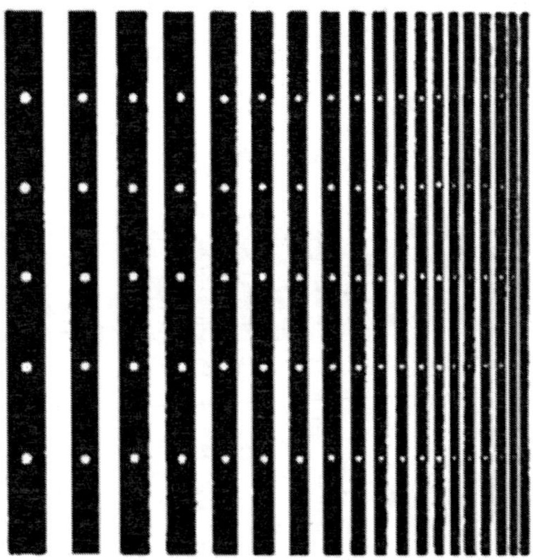

Fig. 139.—The table of a sound-board.

It is necessary to arrange matters so that the pressing down of one key may make all five of the pipes belonging to it speak, or only four, three, two, or one, as we may desire. The pipes are mounted in rows on a *sound-board*, which is built up in several layers. At the top is the *upper board*; below it come the *sliders*, one for each stop; and underneath that the *table*. In Fig. 139 we see part of the table from below. Across the under side are fastened parallel bars with spaces (shown black) left between them. Two other bars are fastened across the ends, so that each groove is enclosed by wood at the top and on all sides. The under side of the table has sheets of leather glued or otherwise attached to it in such a manner that no air can leak from one groove to the next. Upper board, sliders, and table are pierced with rows of holes, to permit the passage of wind from the grooves to the pipes. The grooves under the big pipes are wider than those under the small pipes, as they have to pass more air. The bars between the grooves also vary in width according to the weight of the pipes which they have to carry. The sliders can be moved in and out a short distance in the direction of the axis of the rows of pipes. There is one slider under each row. When a slider is in, the holes in it do not correspond with those in the table and upper board, so that no wind can get from the grooves to the rank over that particular slider. Fig. 140 shows the manner in which the sliders are operated by the little knobs (also called stops) projecting from the casing of the organ within convenient reach of the performer's hands. One stop is in, the other drawn out.

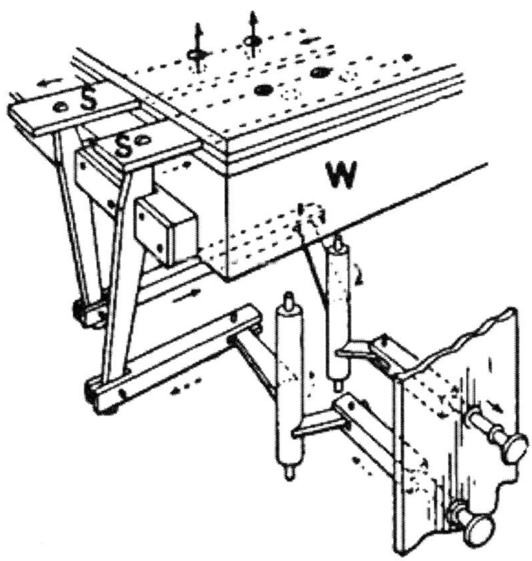

Fig. 140.

In Fig. 141 we see the table, etc., in cross section, with a slider out, putting the pipes of its rank in communication with the grooves. The same diagram shows us in section the little triangular *pallets* which admit air from the *wind-chest* to the grooves; and Fig. 142 gives us an end section of table, sliders, and wind-chest, together with the rods, etc., connecting the key to its pallet. When the key is depressed, the *sticker* (a slight wooden rod) is pushed up. This rocks a *backfall*, or pivoted lever, to which is attached the *pulldown*, a wire penetrating the bottom of the wind-chest to the pallet. As soon as the pallet opens, wind rushes into the groove above through the aperture in the leather bottom, and thence to any one of the pipes of which the slider has been drawn out. (The sliders in Fig. 142 are solid black.) It is evident that if the sound-board is sufficiently deep from back to front, any number of rows of pipes may be placed on it.

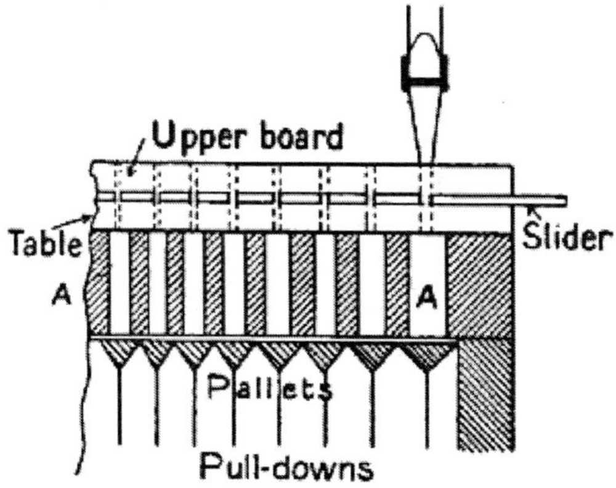

Fig. 141.

PEDALS.

The organ pedals are connected to the pallets by an action similar to that of the keys. The pedal stops are generally of deep tone, 32-foot and 16-foot, as they have to sustain the bass part of the musical harmonies. By means of *couplers* one or more of the keyboard stops may be linked to the pedals.

SEPARATE SOUND-BOARDS.

The keyboard of a very large organ has as many as five *manuals,* or rows of keys. Each manual operates what is practically a separate organ mounted on its own sound-board.

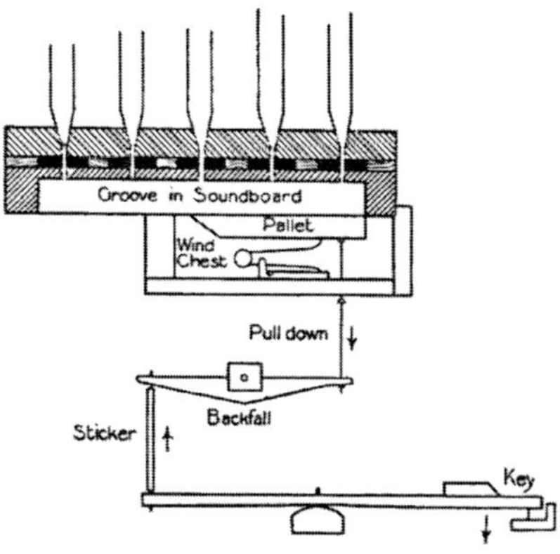

Fig. 142.

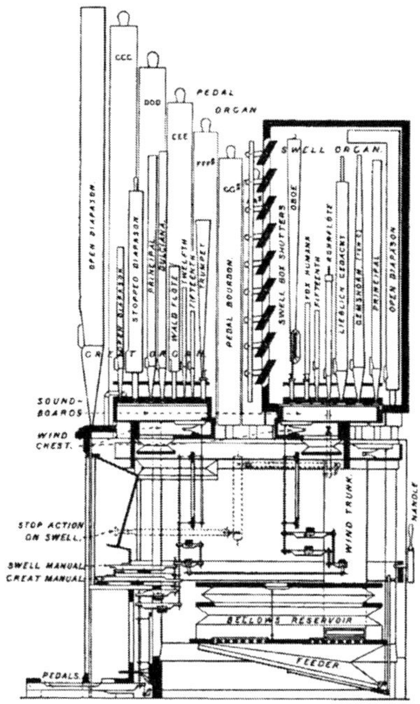

Fig. 143.—General section of a two-manual organ.

The manuals are arranged in steps, each slightly overhanging that below. Taken in order from the top, they are:—(1.) *Echo organ,* of stops of small scale and very soft tone, enclosed in a "swell-box." (2.) *Solo organ,* of stops imitating orchestral instruments. The wonderful "vox humana" stop also belongs to this manual. (3.) *Swell organ,* contained in a swell-box, the front and sides of which have shutters which can be opened and closed by the pressure of the foot on a lever, so as to regulate the amount of sound proceeding from the pipes inside. (4.) *Great organ,* including pipes of powerful tone. (5.) *Choir organ,* of soft, mellow stops, often enclosed in a swell-box. We may add to these the *pedal organ,* which can be coupled to any but the echo manual.

VARIETIES OF STOPS.

We have already remarked that the quality of a stop depends on the shape and construction of the pipe. Some pipes are of wood, others of metal. Some are rectangular, others circular. Some have parallel sides, others taper or expand towards the top. Some are open, others stopped.

The two main classes into which organ pipes may be divided are:—(1.) *Flue* pipes, in which the wind is directed against a lip, as in Fig. 138. (2.) *Reed* pipes—that is, pipes used in combination with a simple device for admitting air into the bottom of the pipe in a series of gusts. Fig. 144 shows a *striking* reed, such as is found in the ordinary motor horn. The elastic metal tongue when at rest stands a very short distance away from the orifice in the reed. When wind is blown through the reed the tongue is sucked against the reed, blocks the current, and springs away again. A *free* reed has a tongue which vibrates in a slot without actually touching the sides. Harmonium and concertina reeds are of this type. In the organ the reed admits air to a pipe of the correct length to sympathize with the rate of the puffs of air which the reed passes. Reed pipes expand towards the top.

TUNING PIPES AND REEDS.

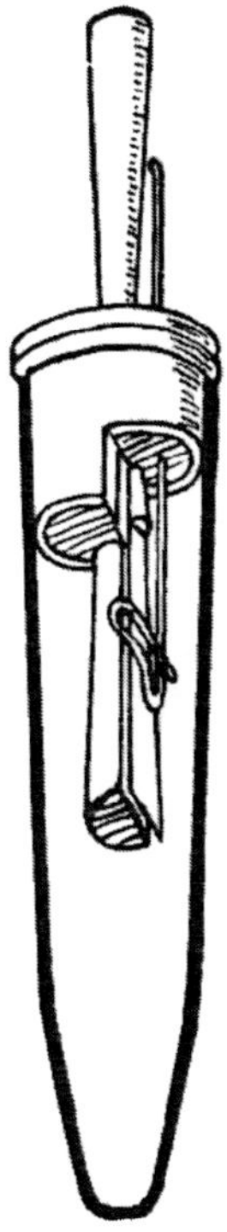

Fig. 144.—A reed pipe.

Pipes are tuned by adjusting their length. The plug at the top of a stopped pipe is pulled out or pushed in a trifle to flatten or sharpen the note respectively. An open pipe, if large, has a tongue cut in the side at the top, which can be pressed inwards or outwards for the purpose of correcting the tone. Small metal pipes are flattened by contracting the tops inwards with a metal cone like a candle-extinguisher placed over the top and tapped; and sharpened by having the top splayed by a cone pushed in point downwards. Reeds of the striking variety (see Fig. 144) have a tuning-wire pressing on the tongue near the fixed end. The end of this wire projects through the casing. By moving it, the length of the vibrating part of the tongue is adjusted to correctness.

BELLOWS.

Different stops require different wind-pressures, ranging from1/10 lb. to 1 lb. to the square inch, the reeds taking the heaviest pressures. There must therefore be as many sets of bellows and wind-chests as there are different pressures wanted. A very large organ consumes immense quantities of air when all the stops are out, and the pumping has to be done by a powerful gas, water, or electric engine. Every bellows has a reservoir (see Fig. 143) above it. The top of this is weighted to give the pressure required. A valve in the top opens automatically as soon as the reservoir has expanded to a certain fixed limit, so that there is no possibility of bursting the leather sides.

Fig. 145.—The keyboard and part of the pneumatic mechanism of the Hereford Cathedral organ.C, composition pedals for pushing out groups of stops;P(at bottom), pedals;P P(at top), pipes carrying compressed air;M, manuals (4); S S, stops.

ELECTRIC AND PNEUMATIC ACTIONS.

We have mentioned in connection with railway signalling that the signalman is sometimes relieved of the hard manual labour of moving signals and points by the employment of electric and pneumatic auxiliaries. The same is true of organs and organists. The touch of the keys has been greatly lightened by making the keys open air-valves or complete electric circuits which actuate the mechanism for pulling down the pallets. The stops, pedals, and couplers also employ "power." Not only are the performer's muscles spared a lot of heavy work when compressed air and electricity aid him, but he is able to have the *console*, or keyboard, far away from the pipes. "From the console, the player, sitting with the singers, or in any desirable part of the choir or chancel, would be able to command the working of the whole of the largest organ situated afar at the western end of the nave; would draw each stop in complete reliance on the sliders and the sound-board fulfilling their office; ... and—marvel of it all—the player, using the swell pedal in his ordinary manner, would obtain crescendo and diminuendo with a more perfect effect than by the old way."[31]

In cathedrals it is no uncommon thing for the different sound-boards to be placed in positions far apart, so that to the uninitiated there may appear to be several independent organs scattered about. Yet all are absolutely under the control of a man who is sitting away from them all, but connected with them by a number of tubes or wires.

The largest organ in the world is that in the Town Hall, Sydney. It has a hundred and twenty-six speaking stops, five manuals, fourteen couplers, and forty-six combination studs. The pipes, about 8,000 in number, range from the enormous 64-foot contra-trombone to some only a fraction of an inch in length. The organ occupies a space 85 feet long and 26 feet deep.

HUMAN REEDS.

The most wonderful of all musical reeds is found in the human throat, in the anatomical part called the *larynx,* situated at the top of the *trachea,* or windpipe.

Slip a piece of rubber tubing over the end of a pipe, allowing an inch or so to project. Take the free part of the tube by two opposite points between the first fingers and thumbs and pull it until the edges are stretched tight. Now blow through it. The wind, forcing its way between the two rubber edges, causes them and the air inside the tube to vibrate, and a musical note results. The more you strain the rubber the higher is the note.

The larynx works on this principle. The windpipe takes the place of the glass pipe; the two vocal cords represent the rubber edges; and the *arytenoid muscles* stand instead of the hands. When contracted, these muscles bring the edges of the cords nearer to one another, stretch the cords, and shorten the cords. A person gifted with a "very good ear" can, it has been calculated, adjust the length of the vocal cords to1/17000th of an inch!

Simultaneously with the adjustment of the cords is effected the adjustment of the length of the windpipe, so that the column of air in it may be of the right length to vibrate in unison. Here again is seen a wonderful provision of nature.

The resonance of the mouth cavity is also of great importance. By altering the shape of the mouth the various harmonics of any fundamental note produced by the larynx are rendered prominent, and so we get the different vocal sounds. Helmholtz has shown that the fundamental tone of any note is represented by the sound *oo*. If the mouth is adjusted to bring out the octave of the fundamental, *o* results. *a* is produced by accentuating the second harmonic, the twelfth; *ee* by developing the second and fourth harmonics; while for *ah* the fifth and seventh must be prominent.

When we whistle we transform the lips into a reed and the mouth into a pipe. The tension of the lips and the shape of the mouth cavity decide the note. The lips are also used as a reed for blowing the flute, piccolo, and all the brass band instruments of the cornet order. In blowing a coach-horn the various harmonics of the fundamental note are brought out by altering the lip tension and the wind pressure. A cornet is practically a coach-horn rolled up into a convenient shape and furnished with three keys, the depression of which puts extra lengths of tubing in connection with the main tube—in fact, makes it longer. One key lowers the fundamental note of the horn half a tone; the second, a full tone; the third, a tone and a half. If the first and third are pressed down together, the note sinks two tones; if the second and third, two and a half tones; and simultaneous depression of all three gives a drop of three tones. The performer thus has seven possible fundamental notes, and several harmonics of each of these at his command; so that by a proper manipulation of the keys he can run up the chromatic scale.

We should add that the cornet tube is an "open" pipe. So is that of the flute. The clarionet is a "stopped" pipe.

[29] It is obvious that in Fig. 136, 2, a pulse will pass fromAtoBand back in one-third the time required for it to pass fromAtoBand back in Fig. 136, *1*.

[30] The science of hearing; from the Greek verb, ἀκούειν, «to hear.»

[31] "Organs and Tuning,".

CHAPTER XVI.
TALKING-MACHINES.

The phonograph—The recorder—The reproducer—The gramophone—The making of records—Cylinder records—Gramophone records.

IN the Patent Office Museum at South Kensington is a curious little piece of machinery—a metal cylinder mounted on a long axle, which has at one end a screw thread chased along it. The screw end rotates in a socket with a thread of equal pitch cut in it. To the other end is attached a handle. On an upright near the cylinder is mounted a sort of drum. The membrane of the drum carries a needle, which, when the membrane is agitated by the air-waves set up by human speech, digs into a sheet of tinfoil wrapped round the cylinder, pressing it into a helical groove turned on the cylinder from end to end. This construction is the first phonograph ever made. Thomas Edison, the "wizard of the West," devised it in 1876; and from this rude parent have descended the beautiful machines which record and reproduce human speech and musical sounds with startling accuracy.

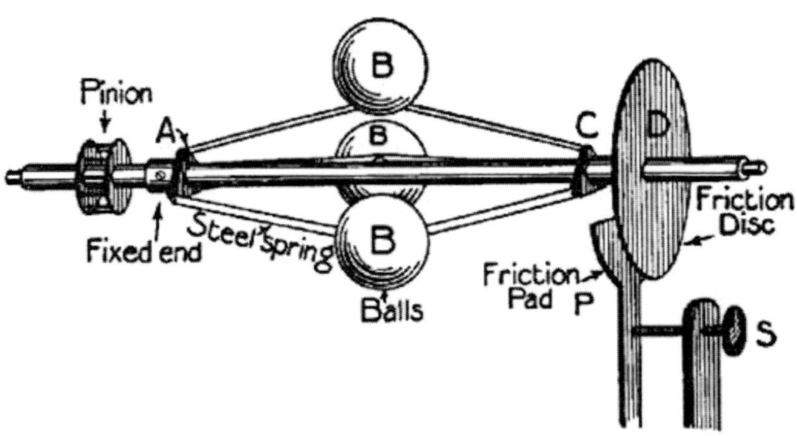

Fig. 146.—The "governor" of a phonograph.

We do not propose to trace here the development of the talking-machine; nor will it be necessary to describe in detail its mechanism, which is probably well known to most readers, or could be mastered in a very short time on personal examination. We will content ourselves with saying that the wax cylinder of the phonograph, or the ebonite disc of the gramophone, is generally rotated by clockwork concealed in the body of the machine. The speed of rotation has to be very carefully governed, in order that the record may revolve under the reproducing point at a uniform speed. The principle of the governor commonly used appears in Fig. 146. The last pinion of the clockwork train is mounted on a shaft carrying two triangular plates,AandC, to which are attached three short lengths of flat steel spring with a heavy ball attached to the centre of each.Ais fixed;Cmoves up the shaft as the balls fly out, and pulls with it the discD, which rubs against the padP(on the end of a spring) and sets up sufficient friction to slow the clockwork. The limit rate is regulated by screwS.

THE PHONOGRAPH.

Though the recording and reproducing apparatus of a phonograph gives very wonderful results, its construction is quite simple. At the same time, it must be borne in mind that an immense amount of experimenting has been devoted to finding out the most suitable materials and forms for the parts.

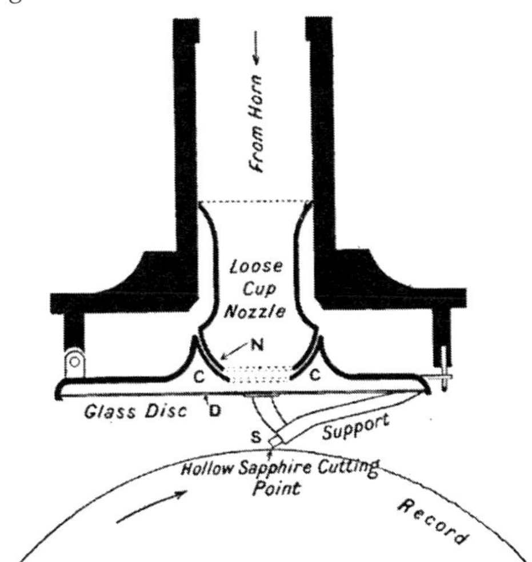

Fig. 147.—Section of an Edison Bell phonograph recorder.

The *recorder* (Fig. 147) is a little circular box about one and a half inches in diameter.[32] From the top a tube leads to the horn. The bottom is a circular plate,C C, hinged at one side. This plate supports a glass disc,D, about1⁄150th of an inch thick, to which is attached the cutting stylus—a tiny sapphire rod with a cup-shaped end having very sharp edges. Sound-waves enter the box through the horn tube; but instead of being allowed to fill the whole box, they are concentrated by the shifting nozzleNon to the centre of the glass disc through the hole inC C. You will notice thatNhas a ball end, andC Ca socket to fitNexactly, so that, thoughC CandNmove up and down very rapidly, they still make perfect contact. The disc is vibrated by the sound-impulses, and drives the cutting point down into the surface of the wax cylinder, turning below it in a clockwork direction. The only dead weight pressing onSis that ofN,C C, and the glass diaphragm.

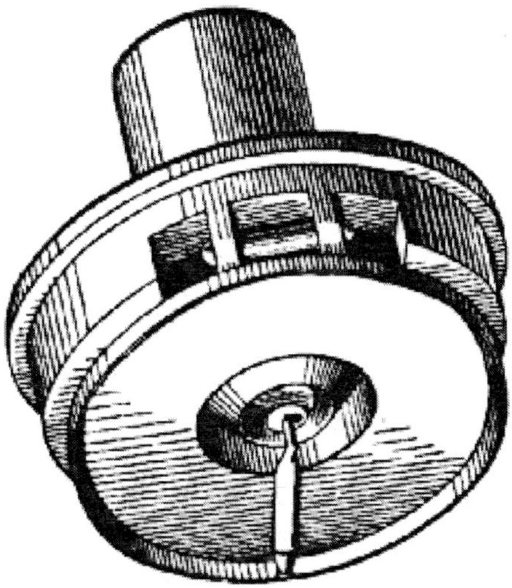

Fig. 148.—Perspective view of a phonograph recorder.

As the cylinder revolves, the recorder is shifted continuously along by a leading screw having one hundred or more threads to the inch cut on it, so that it traces a continuous helical groove from one end of the wax cylinder to the other. This groove is really a series of very minute indentations, not exceeding1⁄1000th of an inch in depth.[33] Seen under a microscope, the surface of the record is a succession of hills and valleys, some much larger than others (Fig. 151, *a*). A loud sound causes the stylus to give a vigorous dig, while low sounds scarcely move it at all. The wonderful thing about this sound-recording is, that not only are the fundamental tones of musical notes impressed, but also the harmonics, which enable us to decide at once whether the record is one of a cornet, violin, or banjo performance. Furthermore, if several instruments are playing simultaneously near the recorder's horn, the stylus catches all the different shades of tone of every note of a chord. There are, so to speak, minor hills and valleys cut in the slopes of the main hills and valleys.

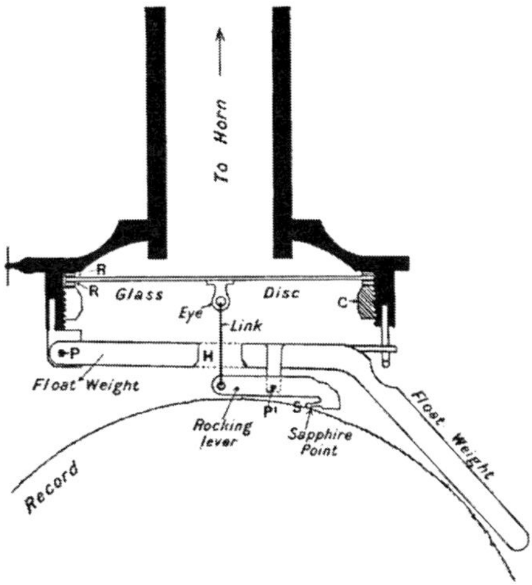

Fig. 149. — Section of the reproducer of an Edison Bell phonograph.

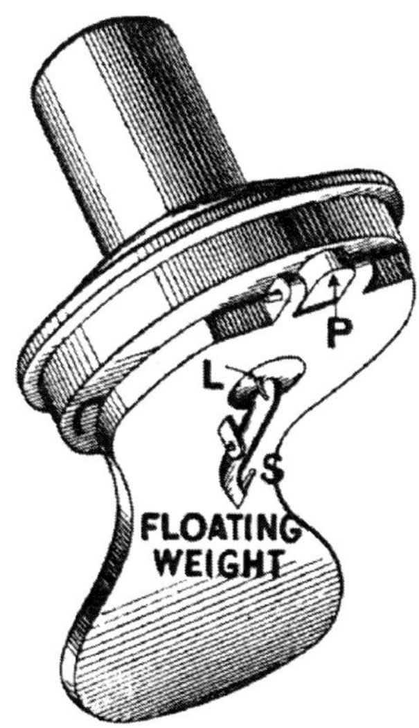

Fig. 150. — Perspective view of a phonograph reproducer.

The *reproducer* (Fig. 149) is somewhat more complicated than the recorder. As before, we have a circular box communicating with the horn of the instrument. A thin glass disc forms a bottom to the box. It is held in position between rubber rings,R R, by a screw collar,C. To the centre is attached a little eye, from which hangs a link,L. Pivoted atPfrom one edge of the box is a *floating weight*, having a circular opening immediately under the eye. The link passes through this to the left end of a tiny lever, which rocks on a pivot projecting from the weight. To the right end of the lever is affixed a sapphire bar, or stylus, with a ball end of a diameter equal to that of the cutting point of the recorder. The floating weight presses the stylus against the record, and also keeps the link between the rocking lever of the glass diaphragm in a state of tension. Every blow given to the stylus is therefore transmitted by the link to the diaphragm, which vibrates and sends an air-impulse into the horn. As the impulses are given at the same rate as those which agitated the diaphragm of the recorder, the sounds which they represent are accurately reproduced, even to the harmonics of a musical note.

THE GRAMOPHONE.

This effects the same purpose as the phonograph, but in a somewhat different manner. The phonograph recorder digs vertically downwards into the surface of the record, whereas the stylus of the gramophone wags from side to side and describes a snaky course (Fig. 151 *b*). It makes no difference in talking-machines whether the reproducing stylus be moved sideways or vertically by the record, provided that motion is imparted by it to the diaphragm.

(a)

Fig. 151 a.

###Insert image155.jpg here###

(b)

Fig. 151 b.

###Insert image156.jpg here###

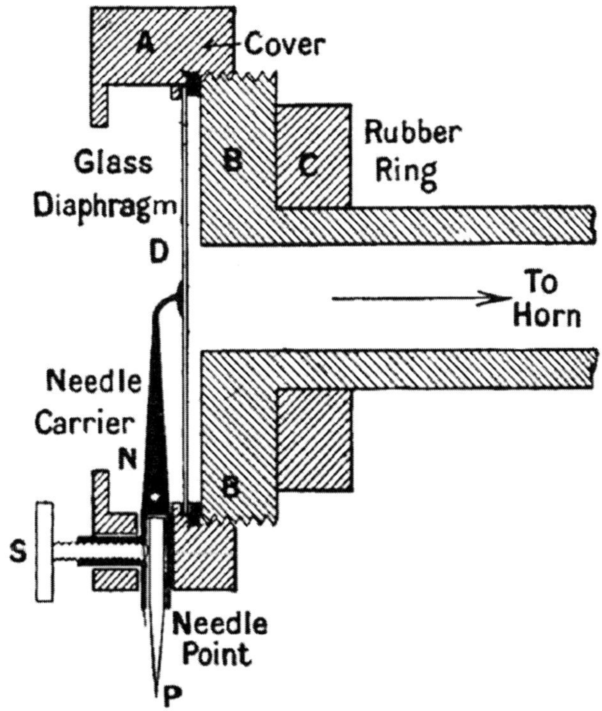

Fig. 151 c.—Section of a gramophone reproducer.

In Fig. 151 *c* the construction of the gramophone reproducer is shown in section.Ais the cover which screws on to the bottomB, and confines the diaphragmDbetween itself and a rubber ring. The portionBis elongated into a tubular shape for connection with the horn, an arm of which slides over the tube and presses against the rubber ringCto make an air-tight joint. The needle-carrierNis attached at its upper end to the centre of the diaphragm. At a point indicated by the white dot a pin passes through it and the cover. The lower end is tubular to accommodate the steel points, which have to be replaced after passing once over a record. A screw,S, working in a socket projecting from the carrier, holds the point fast. The record moves horizontally under the point in a plane perpendicular to the page. The groove being zigzag, the needle vibrates right and left, and rotating the carrier a minute fraction of an inch on the pivot, shakes the glass diaphragm and sends waves of air into the horn.

The gramophone is a reproducing instrument only. The records are made on a special machine, fitted with a device for causing the recorder point to describe a spiral course from the circumference to the centre of the record disc. Some gramophone records have as many as 250 turns to the inch. The total length of the tracing on a ten-inch "concert" record is about 1,000 feet.

THE MAKING OF RECORDS.

For commercial purposes it would not pay to make every record separately in a recording machine. The expense of employing good singers and instrumentalists renders such a method impracticable. All the records we buy are made from moulds, the preparation of which we will now briefly describe.

CYLINDER, OR PHONOGRAPH RECORDS.

First of all, a wax record is made in the ordinary way on a recording machine. After being tested and approved, it is hung vertically and centrally from a rotating table pivoted on a vertical metal spike passing up through the record. On one side of the table is a piece of iron. On each side of the record, and a small distance away, rises a brass rod enclosed in a glass tube. The top of the rods are hooked, so that pieces of gold leaf may be suspended from them. A bell-glass is now placed over the record, table, and rods, and the air is sucked out by a pump. As soon as a good vacuum has been obtained, the current from the secondary circuit of an induction coil is sent into the rods supporting the gold leaves, which are volatilized by the current jumping from one to the other. A magnet, whirled outside the bell-glass, draws round the iron armature on the pivoted table, and consequently revolves the record, on the surface of which a very thin coating of gold is deposited. The record is next placed in an electroplating bath until a copper shell one-sixteenth of an inch thick has formed all over the outside. This is trued up on a lathe and encased in a brass tube. The "master," or original wax record, is removed by cooling it till it contracts sufficiently to fall out of the copper mould, on the inside surface of which are reproduced, in relief, the indentations of the wax "master."

Copies are made from the mould by immersing it in a tank of melted wax. The cold metal chills the wax that touches it, so that the mould soon has a thick waxen lining. The mould and copy are removed from the tank and mounted on a lathe, which shapes and smooths the inside of the record. The record is loosened from the mould by cooling. After inspection for flaws, it is, if found satisfactory, packed in cotton-wool and added to the saleable stock.

Gramophone master records are made on a circular disc of zinc, coated over with a very thin film of acid-proof fat. When the disc is revolved in the recording machine, the sharp stylus cuts through the fat and exposes the zinc beneath. On immersion in a bath of chromic acid the bared surfaces are bitten into, while the unexposed parts remain unaffected. When the etching is considered complete, the plate is carefully cleaned and tested. A negative copper copy is made from it by electrotyping. This constitutes the mould. From it as many as 1,000 copies may be made on ebonite plates by combined pressure and heating.

[32] The Edison Bell phonograph is here referred to.

[33] Some of the sibilant or hissing sounds of the voice are computed to be represented by depressions less than a millionth of an inch in depth. Yet these are reproduced very clearly!

CHAPTER XVII.
WHY THE WIND BLOWS.

Why the wind blows—Land and sea breezes—Light air and moisture—The barometer—The column barometer—The wheel barometer—A very simple barometer—The aneroid barometer—Barometers and weather—The diving-bell—The diving-dress—Air-pumps—Pneumatic tyres—The air-gun—The self-closing door-stop—The action of wind on oblique surfaces—The balloon—The flying-machine.

WHEN a child's rubber ball gets slack through a slight leakage of air, and loses some of its bounce, it is a common practice to hold it for a few minutes in front of the fire till it becomes temporarily taut again. Why does the heat have this effect on the ball? No more air has been forced into the ball. After perusing the chapter on the steam-engine the reader will be able to supply the answer. "Because the molecules of air dash about more vigorously among one another when the air is heated, and by striking the inside of the ball with greater force put it in a state of greater tension."

If we heat an open jar there is no pressure developed, since the air simply expands and flows out of the neck. But the air that remains in the jar, being less in quantity than when it was not yet heated, weighs less, though occupying the same space as before. If we took a very thin bladder and filled it with hot air it would therefore float in colder air, proving that heated air, as we should expect, *tends to rise*. The fire-balloon employs this principle, the air inside the bag being kept artificially warm by a fire burning in some vessel attached below the open neck of the bag.

Now, the sun shines with different degrees of heating power at different parts of the world. Where its effect is greatest the air there is hottest. We will suppose, for the sake of argument, that, at a certain moment, the air envelope all round the globe is of equal temperature. Suddenly the sun shines out and heats the air at a point,A, till it is many degrees warmer than the surrounding air. The heated air expands, rises, and spreads out above the cold air. But, as a given depth of warm air has less weight than an equal depth of cold air, the cold air at once begins to rush towardsBand squeeze the rest of the warm air out. We may therefore picture the atmosphere as made up of a number of colder currents passing along the surface of the earth to replace warm currents rising and spreading over the upper surface of the cold air. A similar circulation takes place in a vessel of heated water.

LAND AND SEA BREEZES.

A breeze which blows from the sea on to the land during the day often reverses its direction during the evening. Why is this? The earth grows hot or cold more rapidly than the sea. When the sun shines hotly, the land warms quickly and heats the air over it, which becomes light, and is displaced by the cooler air over the sea. When the sun sets, the earth and the air over it lose their warmth quickly, while the sea remains at practically the same temperature as before. So the balance is changed, the heavier air now lying over the land. It therefore flows seawards, and drives out the warmer air there.

LIGHT AIR AND MOISTURE.

Light, warm air absorbs moisture. As it cools, the moisture in it condenses. Breathe on a plate, and you notice that a watery film forms on it at once. The cold surface condenses the water suspended in the warm breath. If you wish to dry a damp room you heat it. Moisture then passes from the walls and objects in the room to the atmosphere.

THE BAROMETER.

This property of air is responsible for the changes in weather. Light, moisture-laden air meets cold, dry air, and the sudden cooling forces it to release its moisture, which falls as rain, or floats about as clouds. If only we are able to detect the presence of warm air-strata above us, we ought to be in a position to foretell the weather.

We can judge of the specific gravity of the air in our neighbourhood by means of the barometer, which means "weight-measurer." The normal air-pressure at sea-level on our bodies or any other objects is about 15 lbs. to the square inch—that is to say, if you could imprison and weigh a column of air one inch square in section and of the height of the world's atmospheric envelope, the scale would register 15 lbs. Many years ago (1643) Torricelli, a pupil of Galileo, first calculated the pressure by a very simple experiment. He took a long glass tube sealed at one end, filled it with mercury, and, closing the open end with the thumb, inverted the tube and plunged the open end below the surface of a tank of mercury. On removing his thumb he found that the mercury sank in the tube till the surface of the mercury in the tube was about 30 inches in a vertical direction above the surface of the mercury in the tank. Now, as the upper end was sealed, there must be a vacuum *above* the mercury. What supported the column? The atmosphere. So it was evident that the downward pressure of the mercury exactly counterbalanced the upward pressure of the air. As a mercury column 30 inches high and 1 inch square weighs 15 lbs., the air-pressure on a square inch obviously is the same.

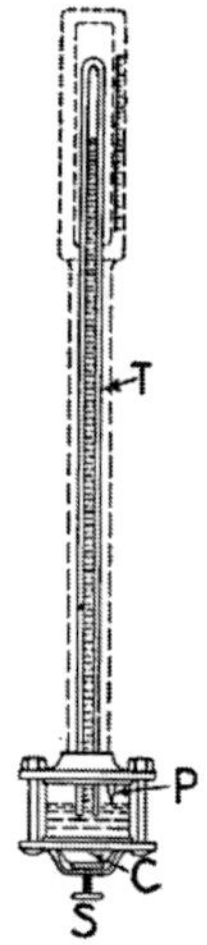

Fig. 152. – A Fortin barometer.

FORTIN'S COLUMN BAROMETER

is a simple Torricellian tube,T, with the lower end submerged in a little glass tank of mercury (Fig. 152). The bottom of this tank is made of washleather. To obtain a "reading" the screwS, pressing on the washleather, is adjusted until the mercury in the tank rises to the tip of the little ivory pointP. The reading is the figure of the scale on the face of the case opposite which the surface of the column stands.

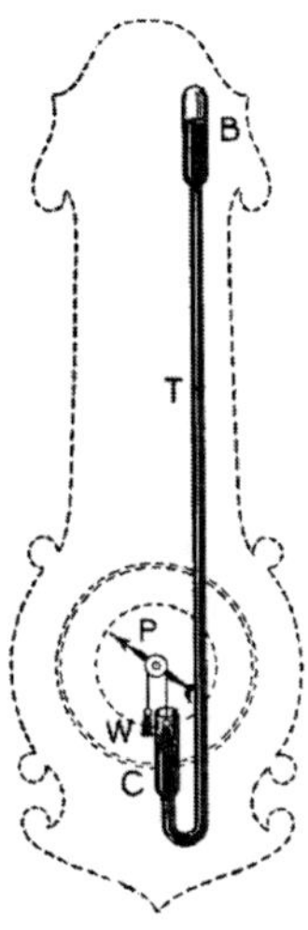

Fig. 153.

THE WHEEL BAROMETER

also employs the mercury column (Fig. 153). The lower end of the tube is turned up and expanded to form a tank,C. The pointerP, which travels round a graduated dial, is mounted on a spindle carrying a pulley, over which passes a string with a weight at each end. The heavier of the weights rests on the top of the mercury. When the atmospheric pressure falls, the mercury inCrises, lifting this weight, and the pointer moves. This form of barometer is not so delicate or reliable as Fortin's, or as the siphon barometer, which has a tube of the same shape as the wheel instrument, but of the same diameter from end to end except for a contraction at the bend. The reading of a siphon is the distance between the two surfaces of the mercury.

A VERY SIMPLE BAROMETER

is made by knocking off the neck of a small bottle, filling the body with water, and hanging it up by a string in the position shown (Fig. 154). When the atmospheric pressure falls, the water at the orifice bulges outwards; when it rises, the water retreats till its surface is slightly concave.

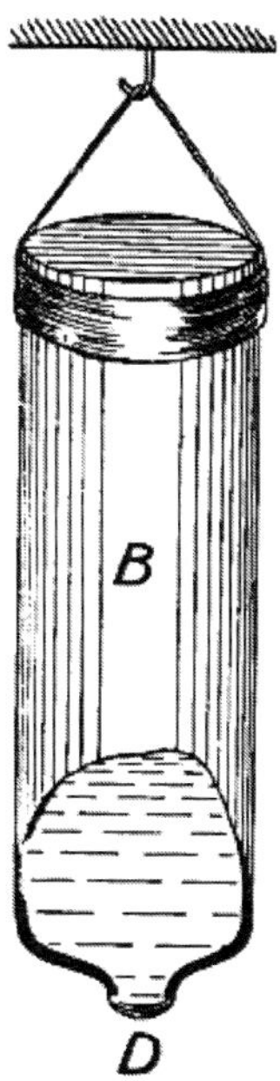

Fig. 154.

THE ANEROID BAROMETER.

On account of their size and weight, and the comparative difficulty of transporting them without derangement of the mercury column, column barometers are not so generally used as the aneroid variety. Aneroid means "without moisture," and in this particular connection signifies that no liquid is used in the construction of the barometer.

Fig. 155 shows an aneroid in detail. The most noticeable feature is the vacuum chamber,V C, a circular box which has a top and bottom of corrugated but thin and elastic metal. Sections of the box are shown in Figs. 156, 157. It is attached at the bottom to the base board of the instrument by a screw (Fig. 156). From the top rises a pin,P, with a transverse hole through it to accommodate the pinK E, which has a triangular section, and stands on one edge.

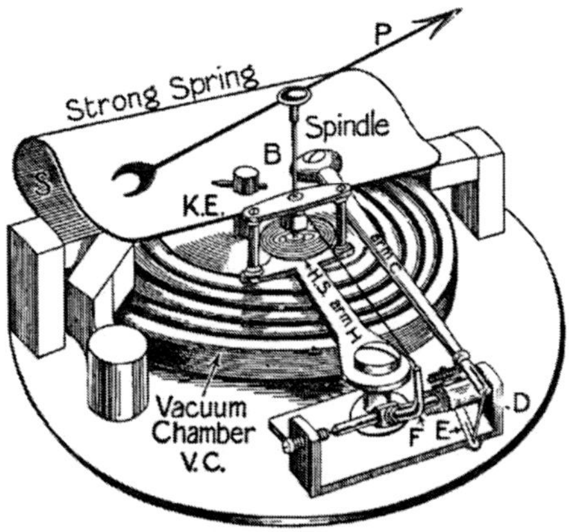

Fig. 155.—An aneroid barometer.

Returning to Fig. 155, we see thatPprojects throughS, a powerful spring of sheet-steel. To this is attached a long arm,C, the free end of which moves a link rotating, through the pinE, a spindle mounted in a frame,D. The spindle moves armF. This pulls on a very minute chain wound round the pointer spindleB, in opposition to a hairspring,H S.Bis mounted on armH, which is quite independent of the rest of the aneroid.

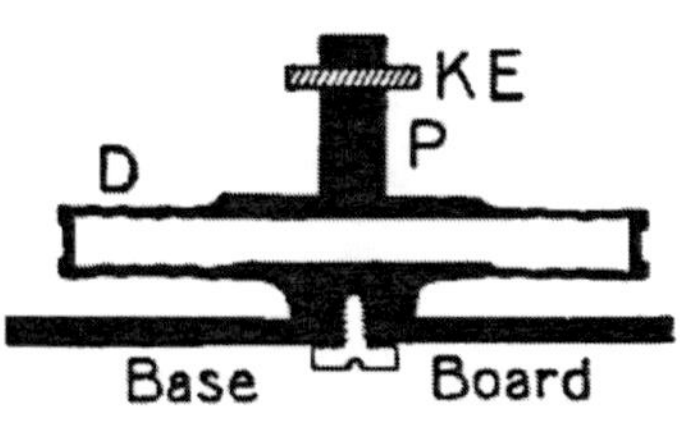

Fig. 156.

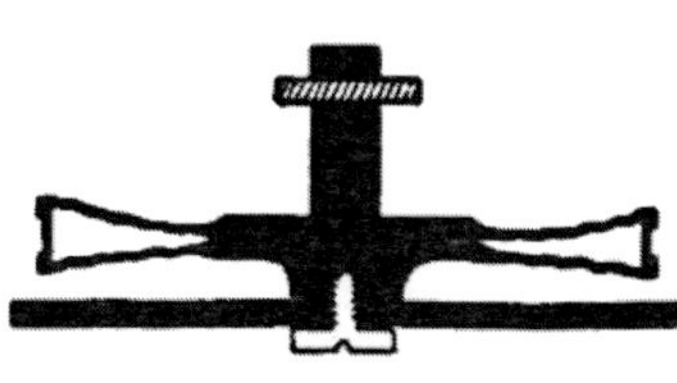

Fig. 157.

The vacuum chamber of an aneroid barometer extended and compressed.

The vacuum chamber is exhausted during manufacture and sealed. It would naturally assume the shape of Fig. 157, but the springS, acting against the atmospheric pressure, pulls it out. As the pressure varies, so does the spring rise or sink; and the slightest movement is transmitted through the multiplying armsC,E,F, to the pointer.

A good aneroid is so delicate that it will register the difference in pressure caused by raising it from the floor to the table, where it has a couple of feet less of air-column resting upon it. An aneroid is therefore a valuable help to mountaineers for determining their altitude above sea-level.

BAROMETERS AND WEATHER.

We may now return to the consideration of forecasting the weather by movements of the barometer. The first thing to keep in mind is, that the instrument is essentially a *weight* recorder. How is weather connected with atmospheric weight?

In England the warm south-west wind generally brings wet weather, the north and east winds fine weather; the reason for this being that the first reaches us after passing over the Atlantic and picking up a quantity of moisture, while the second and third have come overland and deposited their moisture before reaching us.

A sinking of the barometer heralds the approach of heated air—that is, moist air—which on meeting colder air sheds its moisture. So when the mercury falls we expect rain. On the other hand, when the "glass" rises, we know that colder air is coming, and as colder air comes from a dry quarter we anticipate fine weather. It does not follow that the same conditions are found in all parts of the world. In regions which have the ocean to the east or the north, the winds blowing thence would be the rainy winds, while south-westerly winds might bring hot and dry weather.

THE DIVING-BELL.

Water is nearly 773 times as heavy as air. If we submerge a barometer a very little way below the surface of a water tank, we shall at once observe a rise of the mercury column. At a depth of 34 feet the pressure on any submerged object is 15 lbs. to the square inch, in addition to the atmospheric pressure of 15 lbs. per square inch—that is, there would be a 30-lb. *absolute* pressure. As a rule, when speaking of hydraulic pressures, we start with the normal atmospheric pressure as zero, and we will here observe the practice.

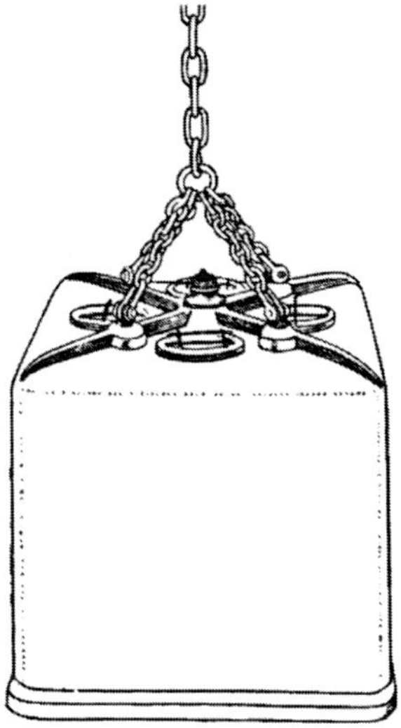

Fig. 158.—A diving bell.

The diving-bell is used to enable people to work under water without having recourse to the diving-dress. A sketch of an ordinary diving-bell is given in Fig. 158. It may be described as a square iron box without a bottom. At the top are links by which it is attached to a lowering chain, and windows, protected by grids; also a nozzle for the air-tube.

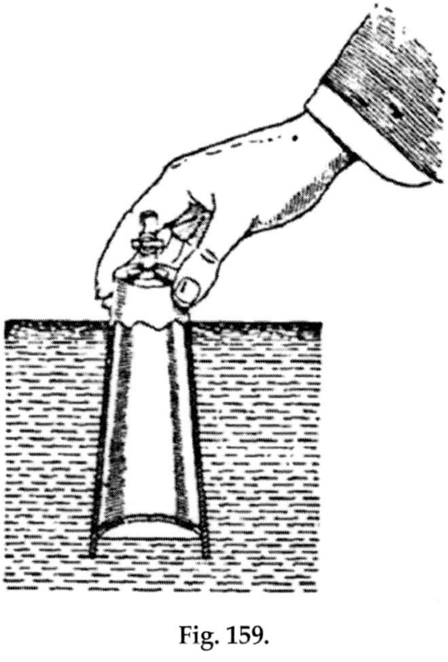

Fig. 159.

A simple model bell (Fig. 159) is easily made out of a glass tumbler which has had a tap fitted in a hole drilled through the bottom. We turn off the tap and plunge the glass into a vessel of water. The water rises a certain way up the interior, until the air within has been compressed to a pressure equal to that of the water at the level of the surface inside. The further the tumbler is lowered, the higher does the water rise inside it.

Evidently men could not work in a diving-bell which is invaded thus by water. It is imperative to keep the water at bay. This we can do by attaching a tube to the tap (Fig. 160) and blowing into the tumbler till the air-pressure exceeds that of the water, which is shown by bubbles rising to the surface. The diving-bell therefore has attached to it a hose through which air is forced by pumps from the atmosphere above, at a pressure sufficient to keep the water out of the bell. This pumping of air also maintains a fresh supply of oxygen for the workers.

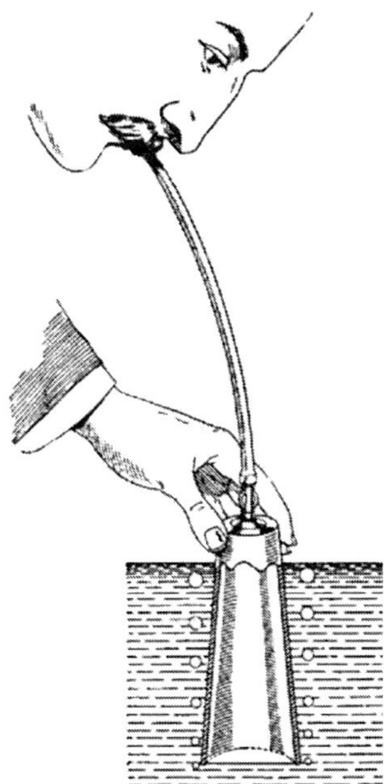

Fig. 160.

Inside the bell is tackle for grappling any object that has to be moved, such as a heavy stone block. The diving-bell is used mostly for laying submarine masonry. "The bell, slung either from a crane on the masonry already built above sea-level, or from a specially fitted barge, comes into action. The block is lowered by its own crane on to the bottom. The bell descends upon it, and the crew seize it with tackle suspended inside the bell. Instructions are sent up as to the direction in which the bell should be moved with its burden, and as soon as the exact spot has been reached the signal for lowering is given, and the stone settles on to the cement laid ready for it."[34]

For many purposes it is necessary that the worker should have more freedom of action than is possible when he is cooped up inside an iron box. Hence the invention of the

DIVING-DRESS,

which consists of two main parts, the helmet and the dress proper. The helmet (Fig. 161) is made of copper. A breastplate,B, shaped to fit the shoulders, has at the neck a segmental screw bayonet-joint. The headpiece is fitted with a corresponding screw, which can be attached or removed by one-eighth of a turn. The neck edge of the dress, which is made in one piece, legs, arms, body and all, is attached to the breastplate by means of the plateP1, screwed down tightly on it by the wing-nutsN N, the bolts of which pass through the breastplate. Air enters the helmet through a valve situated at the back, and is led through tubes along the inside to the front. This valve closes automatically if any accident cuts off the air supply, and encloses sufficient air in the dress to allow the diver to regain the surface. The outlet valveO Vcan be adjusted by the diver to maintain any pressure. At the sides of the headpiece are two hooks,H, over which pass the cords connecting the heavy lead weights of 40 lbs. each hanging on the diver's breast and back. These weights are also attached to the knobsK K. A pair of boots, having 17 lbs. of lead each in the soles, complete the dress. Three glazed windows are placed in the headpiece, that in the front,R W, being removable, so that the diver may gain free access to the air when he is above water without being obliged to take off the helmet.

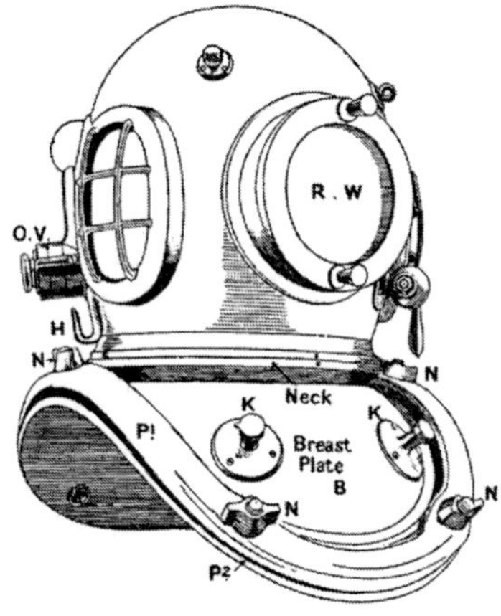

Fig. 161.—A diver's helmet.

By means of telephone wires built into the life-line (which passes under the diver's arms and is used for lowering and hoisting) easy communication is established between the diver and his attendants above. The transmitter of the telephone is placed inside the helmet between the front and a side window, the receiver and the button of an electric bell in the crown. This last he can press by raising his head. The life-line sometimes also includes the wires for an electric lamp (Fig. 162) used by the diver at depths to which daylight cannot penetrate.

The pressure on a diver's body increases in the ratio of 4⅓ lbs. per square inch for every 10 feet that he descends. The ordinary working limit is about 150 feet, though "old hands" are able to stand greater pressures. The record is held by one James Hooper, who, when removing the cargo of the *Cape Horn* sunk off the South American coast, made seven descents of 201 feet, one of which lasted for forty-two minutes.

Fig. 162.—Diver's electric lamp.

A sketch is given (Fig. 163) of divers working below water with pneumatic tools, fed from above with high-pressure air. Owing to his buoyancy a diver has little depressing or pushing power, and he cannot bore a hole in a post with an auger unless he is able to rest his back against some firm object, or is roped to the post. Pneumatic chipping tools merely require holding to their work, their weight offering sufficient resistance to the very rapid blows which they make.

Fig. 163.—Divers at work below water with pneumatic tools.

AIR-PUMPS.

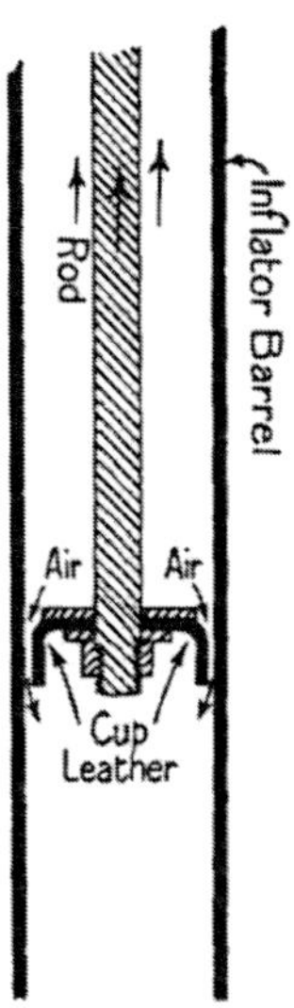

Fig. 164.

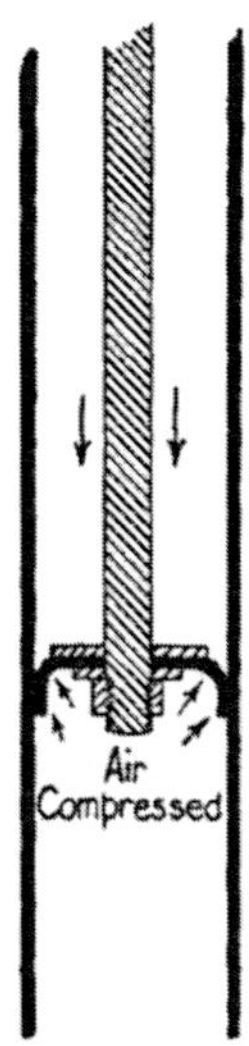

Fig. 165.

Mention having been made of the air-pump, we append diagrams (Figs. 164, 165) of the simplest form of air-pump, the cycle tyre inflator. The piston is composed of two circular plates of smaller diameter than the barrel, holding between them a cup leather. During the upstroke the cup collapses inwards and allows air to pass by it. On the downstroke (Fig. 165) the edges of the cup expand against the barrel, preventing the passage of air round the piston. A double-action air-pump requires a long, well-fitting piston with a cup on each side of it, and the addition of extra valves to the barrel, as the cups under these circumstances cannot act as valves.

PNEUMATIC TYRES.

Fig. 166.

Fig. 167.

The action of the pneumatic tyre in reducing vibration and increasing the speed of a vehicle is explained by Figs. 166, 167. When the tyre encounters an obstacle, such as a large stone, it laps over it (Fig. 166), and while supporting the weight on the wheel, reduces the deflection of the direction of movement. When an iron-tyred wheel meets a similar obstacle it has to rise right over it, often jumping a considerable distance into the air. The resultant motions of the wheel are indicated in each case by an arrow. Every change of direction means a loss of forward velocity, the loss increasing with the violence and extent of the change. The pneumatic tyre also scores because, on account of its elasticity, it gives a "kick off" against the obstacle, which compensates for the resistance during compression.

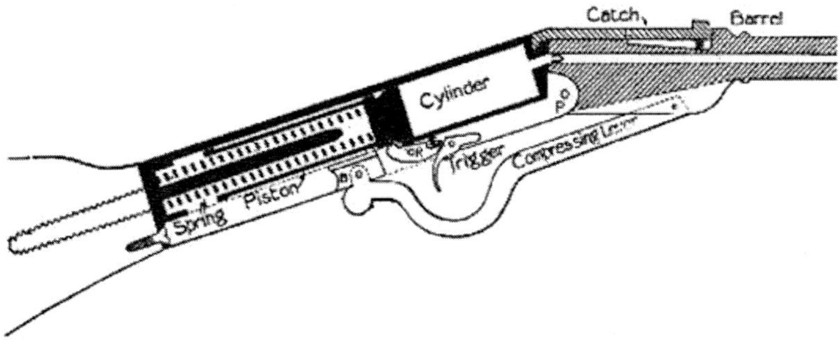

Fig. 168. — Section of the mechanism of an air-gun.

THE AIR-GUN.

This may be described as a valveless air-pump. Fig. 168 is a section of a "Gem" air-gun, with the mechanism set ready for firing. In the stock of the gun is the *cylinder*, in which an accurately fitting and hollow *piston* moves. A powerful helical spring, turned out of a solid bar of steel, is compressed between the inside end of the piston and the upper end of the butt. To set the gun, the *catch* is pressed down so that its hooked end disengages from the stock, and the barrel is bent downwards on pivotP. This slides the lower end of the *compressing lever* towards the butt, and a projection on the guideB, working in a groove, takes the piston with it. When the spring has been fully compressed, the triangular tip of the rocking camRengages with a groove in the piston's head, and prevents recoil when the barrel is returned to its original position. On pulling the trigger, the piston is released and flies up the cylinder with great force, and the air in the cylinder is compressed and driven through the bore of the barrel, blocked by the leaden slug, to which the whole energy of the expanding spring is transmitted through the elastic medium of the air.

There are several other good types of air-gun, all of which employ the principles described above.

THE SELF-CLOSING DOOR-STOP

is another interesting pneumatic device. It consists of a cylinder with an air-tight piston, and a piston rod working through a cover at one end. The other end of the cylinder is pivoted to the door frame. When the door is opened the piston compresses a spring in the cylinder, and air is admitted past a cup leather on the piston to the upper part of the cylinder. This air is confined by the cup leather when the door is released, and escapes slowly through a leak, allowing the spring to regain its shape slowly, and by the agency of the piston rod to close the door.

THE ACTION OF WIND ON OBLIQUE SURFACES.

Why does a kite rise? Why does a boat sail across the wind? We can supply an answer almost instinctively in both cases, "Because the wind pushes the kite or sail aside." It will, however, be worth while to look for a more scientific answer. The kite cannot travel in the direction of the wind because it is confined by a string. But the face is so attached to the string that it inclines at an angle to the direction of the wind. Now, when a force meets an inclined surface which it cannot carry along with it, but which is free to travel in another direction, the force may be regarded as resolving itself into *two* forces, coming from each side of the original line. These are called the *component* forces.

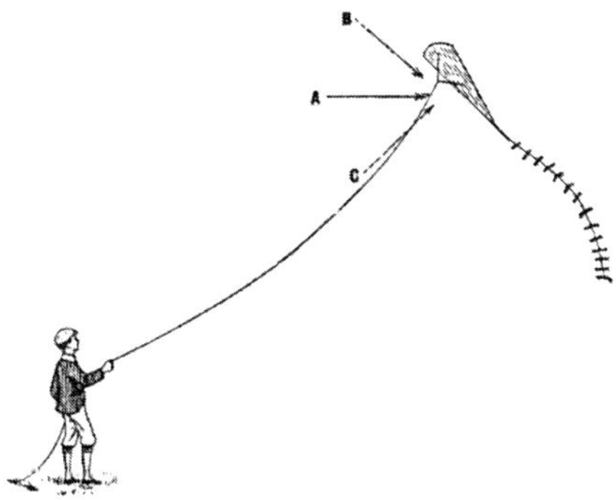

Fig. 169.

To explain this we give a simple sketch of a kite in the act of flying (Fig. 169). The wind is blowing in the direction of the solid arrowA. The oblique surface of the kite resolves its force into the two components indicated by the dotted arrowsBandC. Of theseConly has lifting power to overcome the force of gravity. The kite assumes a position in which forceCand gravity counterbalance one another.

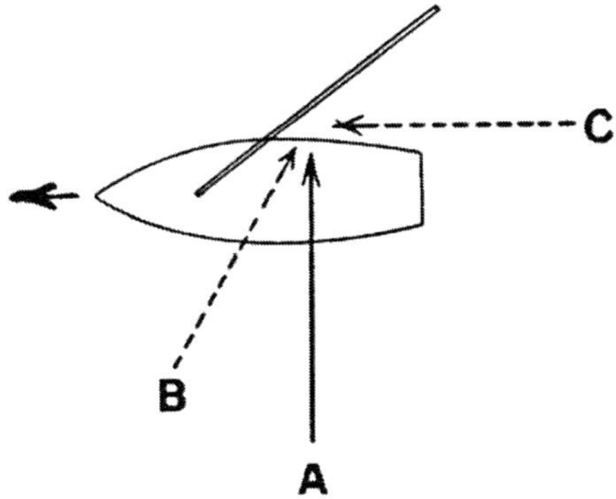

Fig. 170.

A boat sailing across the wind is acted on in a similar manner (Fig. 170). The wind strikes the sail obliquely, and would thrust it to leeward were it not for the opposition of the water. The forceAis resolved into forcesBandC, of whichCpropels the boat on the line of its axis. The boat can be made to sail even "up" the wind, her head being brought round until a point is reached at which the forceBon the boat, masts, etc., overcomes the forceC. The capability of a boat for sailing up wind depends on her "lines" and the amount of surface she offers to the wind.

THE BALLOON

is a pear-shaped bag—usually made of silk—filled with some gas lighter than air. The tendency of a heavier medium to displace a lighter drives the gas upwards, and with it the bag and the wicker-work car attached to a network encasing the bag. The tapering neck at the lower end is open, to permit the free escape of gas as the atmospheric pressure outside diminishes with increasing elevation. At the top of the bag is a wooden valve opening inwards, which can be drawn down by a rope passing up to it through the neck whenever the aeronaut wishes to let gas escape for a descent. He is able to cause a very rapid escape by pulling another cord depending from a "ripping piece" near the top of the bag. In case of emergency this is torn away bodily, leaving a large hole. The ballast (usually sand) carried enables him to maintain a state of equilibrium between the upward pull of the gas and the downward pull of gravity. To sink he lets out gas, to rise he throws out ballast; and this process can be repeated until the ballast is exhausted. The greatest height ever attained by aeronauts is the $7\frac{1}{4}$ miles, or 37,000 feet, of Messrs. Glaisher and Coxwell on September 5, 1862. The ascent nearly cost them their lives, for at an elevation of about 30,000 feet they were partly paralyzed by the rarefaction of the air, and had not Mr. Coxwell been able to pull the valve rope with his teeth and cause a descent, both would have died from want of air.

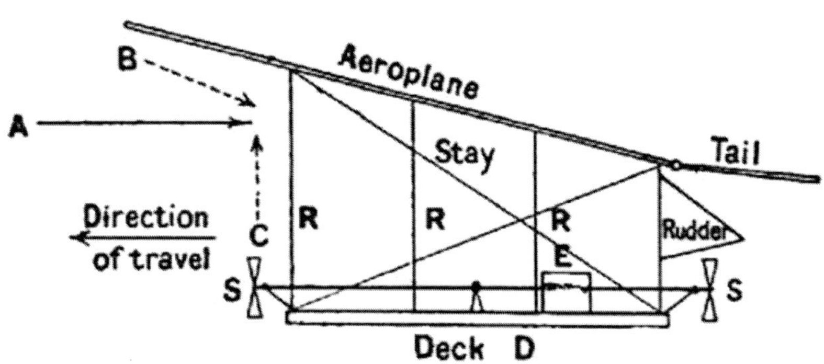

Fig. 171.

The *flying-machine,* which scientific engineers have so long been trying to produce, will probably be quite independent of balloons, and will depend for its ascensive powers on the action of air on oblique surfaces. Sir Hiram Maxim's experimental air-ship embodied the principles shown by Fig. 171. On a deck was mounted an engine,E, extremely powerful for its weight. This drove large propellers,S S. Large aeroplanes, of canvas stretched over light frameworks, were set up overhead, the forward end somewhat higher than the rear. The machine was run on rails so arranged as to prevent it rising. Unfortunately an accident happened at the first trial and destroyed the machine.

In actual flight it would be necessary to have a vertical rudder for altering the horizontal direction, and a horizontal "tail" for steering up or down. The principle of an aeroplane is that of the kite, with this difference, that, instead of moving air striking a captive body, a moving body is propelled against more or less stationary air. The resolution of forces is shown by the arrows as before.

Up to the present time no practical flying-machine has appeared. But experimenters are hard at work examining the conditions which must be fulfilled to enable man to claim the "dominion of the air."

[34] The "Romance of Modern Mechanism,"

CHAPTER XVIII.
HYDRAULIC MACHINERY.

The siphon—The bucket pump—The force-pump—The most marvellous pump—The blood channels—The course of the blood—The hydraulic press—Household water-supply fittings—The ball-cock—The water-meter—Water-supply systems—The household filter—Gas traps—Water engines—The cream separator—The "hydro."

IN the last chapter we saw that the pressure of the atmosphere is 15 lbs. to the square inch. Suppose that to a very long tube having a sectional area of one square inch we fit an air-tight piston (Fig. 172), and place the lower end of the tube in a vessel of water. On raising the piston a vacuum would be created in the tube, did not the pressure of the atmosphere force water up into the tube behind the piston. The water would continue to rise until it reached a point 34 feet perpendicularly above the level of the water in the vessel. The column would then weigh 15 lbs., and exactly counterbalance the atmospheric pressure; so that a further raising of the piston would not raise the water any farther. At sea-level, therefore, the *lifting* power of a pump by suction is limited to 34 feet. On the top of a lofty mountain, where the air-pressure is less, the height of the column would be diminished—in fact, be proportional to the pressure.

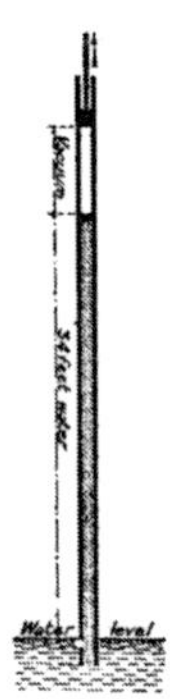

Fig. 172.

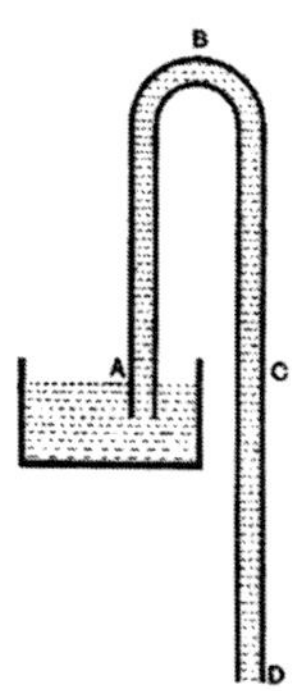

Fig. 173.

THE SIPHON

is an interesting application of the principle of suction. By its own weight water may be made to lift water through a height not exceeding 34 feet. This is explained by Fig. 173. The siphon pipe,A B C D, is in the first instance filled by suction. The weight of the water betweenAandBcounter-balances that betweenBandC. But the columnC Dhangs, as it were, to the heels ofB C, and draws it down. Or, to put it otherwise, the columnB D, being heavier than the columnB A, draws it over the topmost point of the siphon. Any parting between the columns, provided thatB Adoes not exceed 34 feet, is impossible, as the pressure of the atmosphere on the mouth ofB Ais sufficient to prevent the formation of a vacuum.

THE BUCKET PUMP.

We may now pass to the commonest form of pump used in houses, stables, gardens, etc. (Fig. 174). The piston has a large hole through it, over the top of which a valve is hinged. At the bottom of the barrel is a second valve, also opening upwards, seated on the top of the supply pipe. In sketch (*a*) the first upstroke is in progress. A vacuum forms under the piston, or plunger, and water rises up the barrel to fill it. The next diagram (*b*) shows the first downstroke. The plunger valve now opens and allows water to rise above the piston, while the lower closes under the pressure of the water above and the pull of that below. During the second upstroke (*c*) the water above the piston is raised until it overflows through the spout, while a fresh supply is being sucked in below.

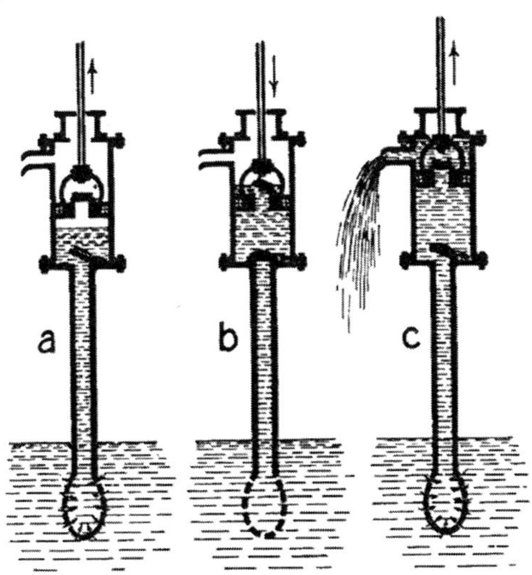

Fig. 174.

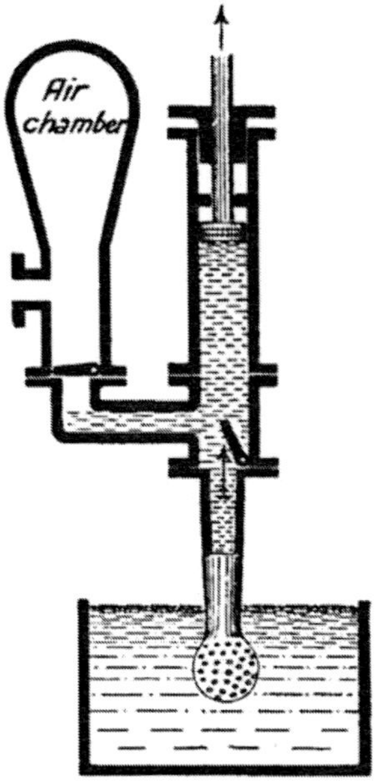

Fig. 175. Force-pump; suction stroke.

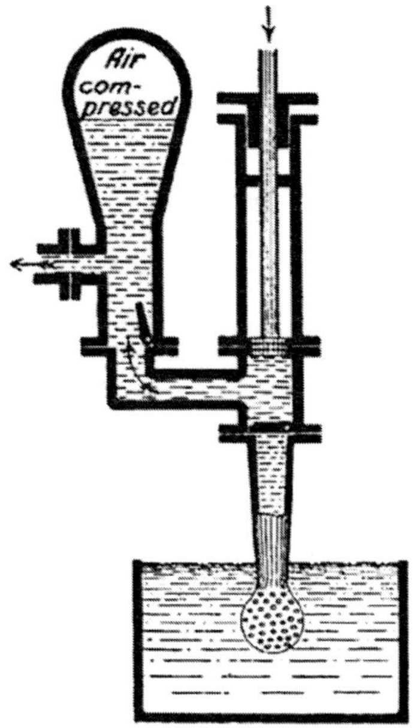

Fig. 176. Force-pump; delivery stroke.

For driving water to levels above that of the pump a somewhat different arrangement is required. One type of force-pump is shown in Figs. 175, 176. The piston now is solid, and the upper valve is situated in the delivery pipe. During an upstroke this closes, and the other opens; the reverse happening during a downstroke. An air-chamber is generally fitted to the delivery pipe when water is to be lifted to great heights or under high pressure. At each delivery stroke the air in the chamber is compressed, absorbing some of the shock given to the water in the pipe by the water coming from the pump; and its expansion during the next suction stroke forces the water gradually up the pipe. The air-chamber is a very prominent feature of the fire-engine.

A *double-action* force-pump is seen in Fig. 177, making an upward stroke. Both sides of the piston are here utilized, and the piston rod works through a water-tight stuffing-box. The action of the pump will be easily understood from the diagram.

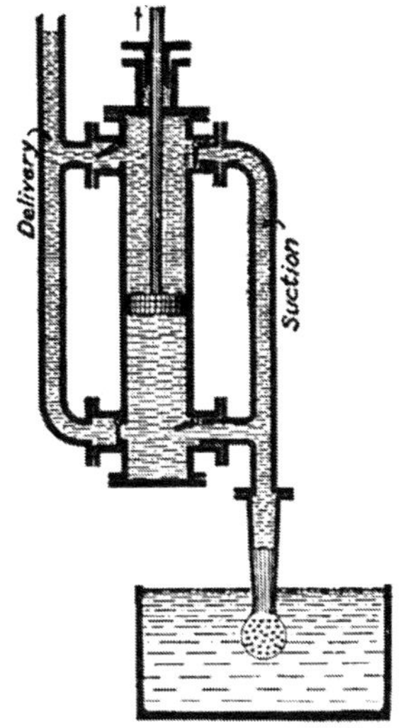

Fig. 177.

THE MOST MARVELLOUS PUMP

known is the *heart*. We give in Fig. 178 a diagrammatic sketch of the system of blood circulation in the human body, showing the heart, the arteries, and the veins, big and little. The body is supposed to be facing the reader, so that the left lung, etc., is to his right.

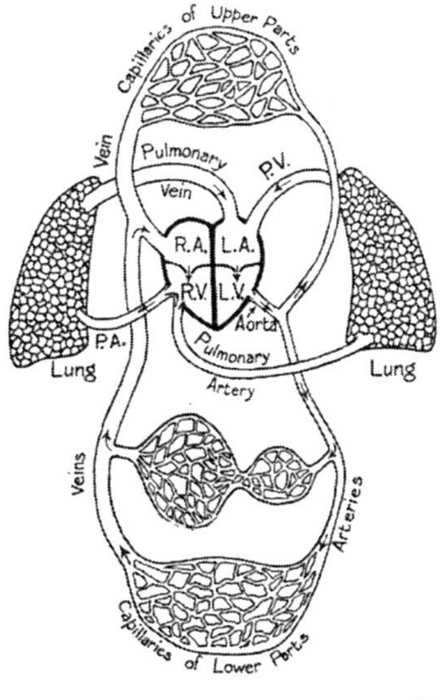

Fig. 178. – A diagrammatic representation of the circulatory system of the blood.

The heart, which forces the blood through the body, is a large muscle (of about the size of the clenched fist) with four cavities. These are respectively known as the right and left *auricles,* and the right and left *ventricles*. They are arranged in two pairs, the auricle uppermost, separated by a fleshy partition. Between each auricle and its ventricle is a valve, which consists of strong membranous flaps, with loose edges turned downwards. The left-side valve is the *mitral* valve, that between the right auricle and ventricle the *tricuspid* valve. The edges of the valves fall together when the heart contracts, and prevent the passage of blood. Each ventricle has a second valve through which it ejects the blood. (That of the right ventricle has been shown double for the sake of convenience.)

The action of the heart is this:—The auricles and ventricles expand; blood rushes into the auricles from the channels supplying them, and distends them and the ventricles; the auricles contract and fill the ventricles below quite full (there are no valves above the auricles, but the force of contraction is not sufficient to return the blood to the veins); the ventricles contract; the mitral and tricuspid valves close; the valves leading to the arteries open; blood is forced out of the ventricles.

THE BLOOD CHANNELS

are of two kinds—(1) The *arteries,* which lead the blood into the circulatory system; (2) the *veins,* which lead the blood back to the heart. The arteries divide up into branches, and these again divide into smaller and smaller arteries. The smallest, termed *capillaries* (Latin, *capillus,* a hair), are minute tubes having an average diameter of1/3000th of an inch. These permeate every part of the body. The capillary arteries lead into the smallest veins, which unite to form larger and larger veins, until what we may call the main streams are reached. Through these the blood flows to the heart.

There are three main points of difference between arteries and veins. In the first place, the larger arteries have thick elastic walls, and maintain their shape even when empty. This elasticity performs the function of the air-chamber of the force-pump. When the ventricles contract, driving blood into the arteries, the walls of the latter expand, and their contraction pushes the blood steadily forward without shock. The capillaries have very thin walls, so that fluids pass through them to and from the body, feeding it and taking out waste matter. The veins are all thin-walled, and collapse when empty. Secondly, most veins are furnished with valves, which prevent blood flowing the wrong way. These are similar in principle to those of the heart. Arteries have no valves. Thirdly, arteries are generally deeply set, while many of the veins run near the surface of the body. Those on the front of the arm are specially visible. Place your thumb on them and run it along towards the wrist, and you will notice that the veins distend owing to the closing of the valves just mentioned.

Arterial blood is *red*, and comes out from a cut in gulps, on account of the contraction of the elastic walls. If you cut a vein, *blue* blood issues in a steady stream. The change of colour is caused by the loss of oxygen during the passage of the blood through the capillaries, and the absorption of carbon dioxide from the tissues.

The *lungs* are two of the great purifiers of the blood. As it circulates through them, it gives up the carbon dioxide which it has absorbed, and receives pure oxygen in exchange. If the air of a room is "foul," the blood does not get the proper amount of oxygen. For this reason it is advisable for us to keep the windows of our rooms open as much as possible both day and night. Fatigue is caused by the accumulation of carbon dioxide and other impurities in the blood. When we run, the heart pumps blood through the lungs faster than they can purify it, and eventually our muscles become poisoned to such an extent that we have to stop from sheer exhaustion.

THE COURSE OF THE BLOOD.

It takes rather less than a minute for a drop of blood to circulate from the heart through the whole system and back to the heart.

We may briefly summarize the course of the circulation of the blood thus:—It is expelled from the left ventricle into the *aorta* and the main arteries, whence it passes into the smaller arteries, and thence into the capillaries of the brain, stomach, kidneys, etc. It here imparts oxygen to the body, and takes in impurities. It then enters the veins, and through them flows back to the right auricle; is driven into the right ventricle; is expelled into the *pulmonary* (lung) *arteries*; enters the lungs, and is purified. It returns to the left auricle through the *pulmonary veins*; enters the left auricle, passes to left ventricle, and so on.

A healthy heart beats from 120 times per minute in a one-year-old infant to 60 per minute in a very aged person. The normal rate for a middle-aged adult is from 80 to 70 beats.

Heart disease signifies the failure of the heart valves to close properly. Blood passes back when the heart contracts, and the circulation is much enfeebled. By listening through a stethoscope the doctor is able to tell whether the valves are in good order. A hissing sound during the beat indicates a leakage past the valves; a thump, or "clack," that they shut completely.

THE HYDRAULIC PRESS.

It is a characteristic of fluids and gases that if pressure be brought to bear on any part of a mass of either class of bodies it is transmitted equally and undiminished in all directions, and acts with the same force on all equal surfaces, at right angles to those surfaces. The great natural philosopher Pascal first formulated this remarkable fact, of which a simple illustration is given in Fig. 179. Two cylinders,AandB, having a bore of one and two inches respectively, are connected by a pipe. Water is poured in, and pistons fitting the cylinders accurately and of equal weight are inserted. On pistonBis placed a load of 10 lbs. To preventArising above the level ofB, it must be loaded proportionately. The area of pistonAis four times that ofB, so that if we lay on it a 40-lb. weight, neither piston will move. The walls of the cylinders and connecting pipe are also pressed outwards in the ratio of 10 lbs. for every part of their interior surface which has an area equal to that of pistonB.

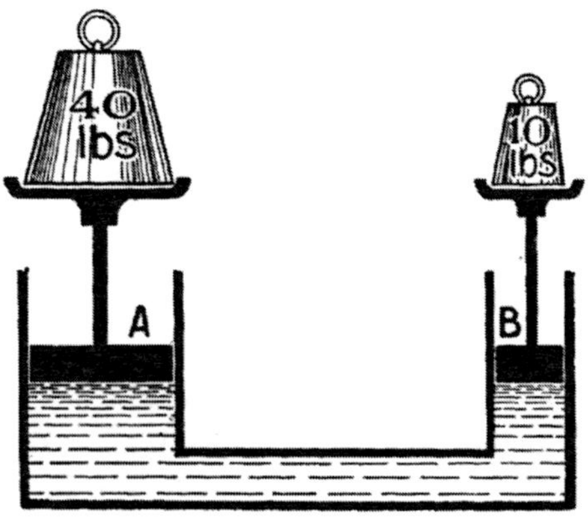

Fig. 179.

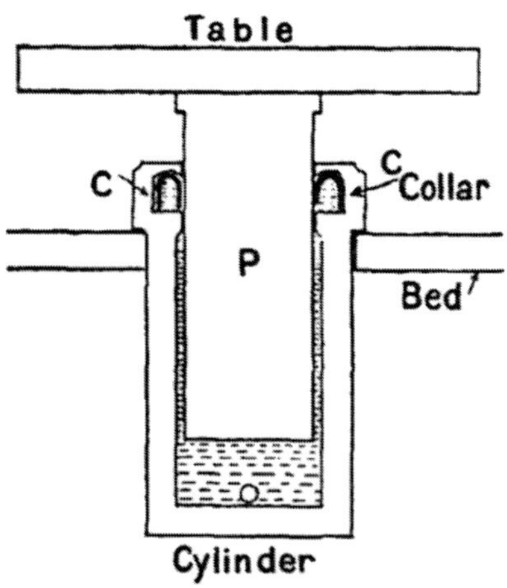

Fig. 180.—The cylinder and ram of a hydraulic press.

The hydraulic press is an application of this law. CylinderBis represented by a force pump of small bore, capable of delivering water at very high pressures (up to 10 tons per square inch). In the place ofAwe have a stout cylinder with a solid plunger,P(Fig. 180), carrying the *table* on which the object to be pressed is placed. Bramah, the inventor of the hydraulic press, experienced great difficulty in preventing the escape of water between the top of the cylinder and the plunger. If a "gland" packing of the type found in steam-cylinders were used, it failed to hold back the water unless it were screwed down so tightly as to jam the plunger. He tried all kinds of expedients without success; and his invention, excellent though it was in principle, seemed doomed to failure, when his foreman, Henry Maudslay,[35] solved the problem in a simple but most masterly manner. He had a recess turned in the neck of the cylinder at the point formerly occupied by the stuffing-box, and into this a leather collar ofU-section (marked solid black in Fig. 180) was placed with its open side downwards. When water reached it, it forced the edges apart, one against the plunger, the other against the walls of the recess, with a degree of tightness proportionate to the pressure. On water being released from the cylinder the collar collapsed, allowing the plunger to sink without friction.

The principle of the hydraulic press is employed in lifts; in machines for bending, drilling, and riveting steel plates, or forcing wheels on or off their axles; for advancing the "boring shield" of a tunnel; and for other purposes too numerous to mention.

HOUSEHOLD WATER-SUPPLY FITTINGS.

Among these, the most used is the tap, or cock. When a house is served by the town or district water supply, the fitting of proper taps on all pipes connected with the supply is stipulated for by the water-works authorities. The old-fashioned "plug" tap is unsuitable for controlling high-pressure water on account of the suddenness with which it checks the flow. Lest the reader should have doubts as to the nature of a plug tap, we may add that it has a tapering cone of metal working in a tapering socket. On the cone being turned till a hole through it is brought into line with the channel of the tap, water passes. A quarter turn closes the tap.

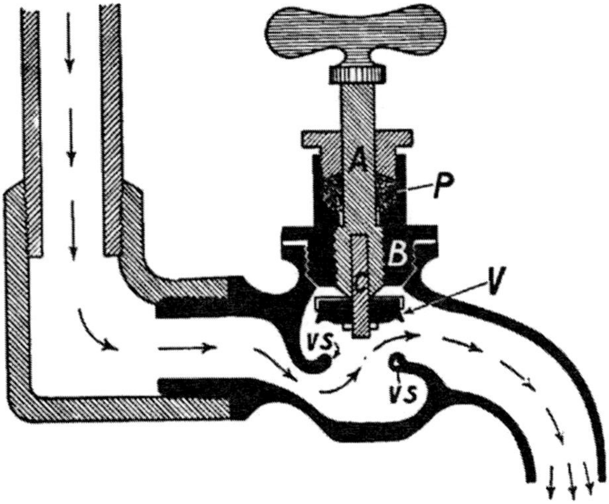

Fig. 181.—A screw-down water cock.

Its place has been taken by the screw-down cock. A very common and effective pattern is shown in Fig. 181. The valveV, with a facing of rubber, leather, or some other sufficiently elastic substance, is attached to a pin,C, which projects upwards into the spindleAof the tap. This spindle has a screw thread on it engaging with a collar,B. When the spindle is turned it rises or falls, allowing the valve to leave its seating,V S, or forcing it down on to it. A packingPin the neck ofBprevents the passage of water round the spindle. To open or close the tap completely is a matter of several turns, which cannot be made fast enough to produce a "water-hammer" in the pipes by suddenly arresting the flow. The reader will easily understand that if water flowing at the rate of several miles an hour is abruptly checked, the shock to the pipes carrying it must be very severe.

THE BALL-COCK

is used to feed a cistern automatically with water, and prevent the water rising too far in the cistern (Fig. 182). Water enters the cistern through a valve, which is opened and closed by a plug faced with rubber. The lower extremity of the plug is flattened, and has a rectangular hole cut in it. Through this passes a lever,L, attached at one end to a hollow copper sphere, and pivoted at the other on the valve casing. This casing is not quite circular in section, for two slots are cast in the circumference to allow water to pass round the plug freely when the valve is open. The buoyancy of the copper sphere is sufficient to force the plug's face up towards its seating as the valve rises, and to cut off the supply entirely when a certain level has been attained. If water is drawn off, the sphere sinks, the valve opens, and the loss is made good.

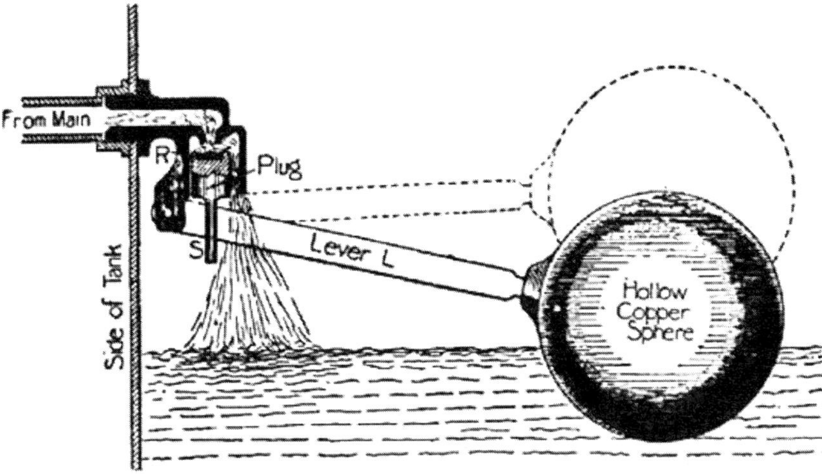

Fig. 182. — An automatic ball-valve.

THE WATER-METER.

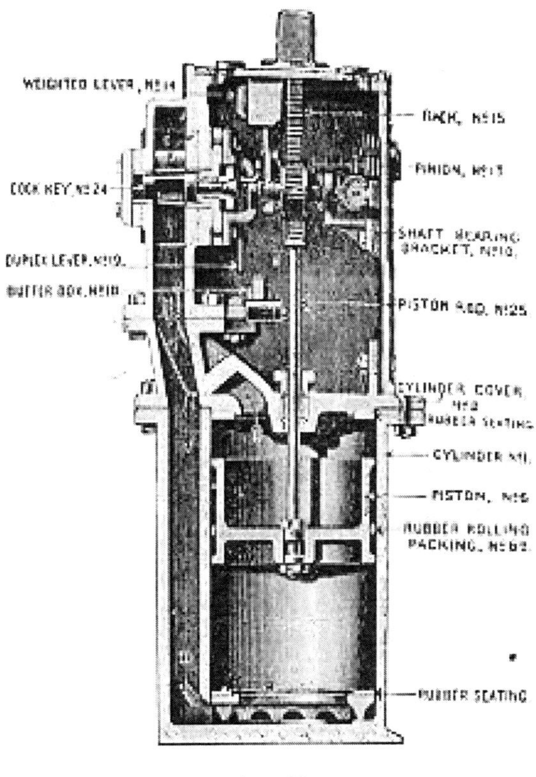

Fig. 183.

Some consumers pay a sum quarterly for the privilege of a water supply, and the water company allows them to use as much as they require. Others, however, prefer to pay a fixed amount for every thousand gallons used. In such cases, a water-meter is required to record the consumption. We append a sectional diagram of Kennedy's patent water-meter (Fig. 183), very widely used. At the bottom is the measuring cylinder, fitted with a piston, (6), which is made to move perfectly water-tight and free from friction by means of a cylindrical ring of india-rubber, rolling between the body of the piston and the internal surface of the cylinder. The piston rod (25), after passing through a stuffing-box in the cylinder cover, is attached to a rack, (15), which gears with a cog, (13), fixed on a shaft. As the piston moves up and down, this cog is turned first in one direction, then in the other. To this shaft is connected the index mechanism (to the right). The cock-key (24) is so constructed that it can put either end of the measuring cylinder in communication with the supply or delivery pipes, if given a quarter turn (see Fig. 184). The weighted lever (14) moves loosely on the pinion shaft through part of a circle. From the pinion project two arms, one on each side of the lever. When the lever has been lifted by one of these past the vertical position, it falls by its own weight on to a buffer-box rest, (18). In doing so, it strikes a projection on the duplex lever (19), which is joined to the cock-key, and gives the latter a quarter turn.

In order to follow the working of the meter, we must keep an eye on Figs. 183 and 184 simultaneously. Water is entering fromA, the supply pipe. It flows through the cock downwards through channelDinto the lower half of the cylinder. The piston rises, driving out the water above it throughCto the delivery pipeB. Just as the piston completes its stroke the weight, raised by the rack and pinion, topples over, and strikes the key-arm, which it sends down till stopped by the buffer-box. The tap is then at right angles to the position shown in Fig. 184, and water is directed fromAdownCinto the top of the cylinder, forcing the piston down, while the water admitted below during the last stroke is forced up the passageD, and out by the outletB. Before the piston has arrived at the bottom of the cylinder, the lifter will have lifted the weighted lever from the buffer-box, and raised it to a vertical position; from there it will have fallen on the right-hand key-arm, and have brought the cock-key to its former position, ready to begin another upward stroke.

Fig. 184.

The *index mechanism* makes allowance for the fact that the bevel-wheel on the pinion shaft has its direction reversed at the beginning of every stroke of the piston. This bevel engages with two others mounted loosely on the little shaft, on which is turned a screw thread to revolve the index counter wheels. Each of these latter bevels actuates the shaft through a ratchet; but while one turns the shaft when rotating in a clockwise direction only, the other engages it when making an anti-clockwise revolution. The result is that the shaft is always turned in the same direction.

WATER-SUPPLY SYSTEMS.

The water for a town or a district supply is got either from wells or from a river. In the former case it may be assumed to be free from impurities. In the latter, there is need for removing all the objectionable and dangerous matter which river water always contains in a greater or less degree. This purification is accomplished by first leading the water into large *settling tanks,* where the suspended matter sinks to the bottom. The water is then drawn off into *filtration beds,* made in the following manner. The bottom is covered with a thick layer of concrete. On this are laid parallel rows of bricks, the rows a small distance apart. Then come a layer of bricks or tiles placed close together; a layer of coarse gravel; a layer of finer gravel; and a thick layer of sand at the top. The sand arrests any solid matter in the water as it percolates to the gravel and drains below. Even the microbes,[36] of microscopic size, are arrested as soon as the film of mud has formed on the top of the sand. Until this film is formed the filter is not in its most efficient condition. Every now and then the bed is drained, the surface mud and sand carefully drained off, and fresh sand put in their place. A good filter bed should not pass more than from two to three gallons per hour for every square foot of surface, and it must therefore have a large area.

It is sometimes necessary to send the water through a succession of beds, arranged in terraces, before it is sufficiently pure for drinking purposes.

THE HOUSEHOLD FILTER.

When there is any doubt as to the wholesomeness of the water supply, a small filter is often used. The microbe-stopper is usually either charcoal, sand, asbestos, or baked clay of some kind. In Fig. 185 we give a section of a Maignen filter.Ris the reservoir for the filtered water;Athe filter case proper;Da conical perforated frame;Ba jacket of asbestos cloth secured top and bottom by asbestos cords toD;Cpowdered carbon, between which and the asbestos is a layer of special chemical filtering medium. A perforated cap,E, covers in the carbon and prevents it being disturbed when water is poured in. The carbon arrests the coarser forms of matter; the asbestos the finer. The asbestos jacket is easily removed and cleansed by heating over a fire.

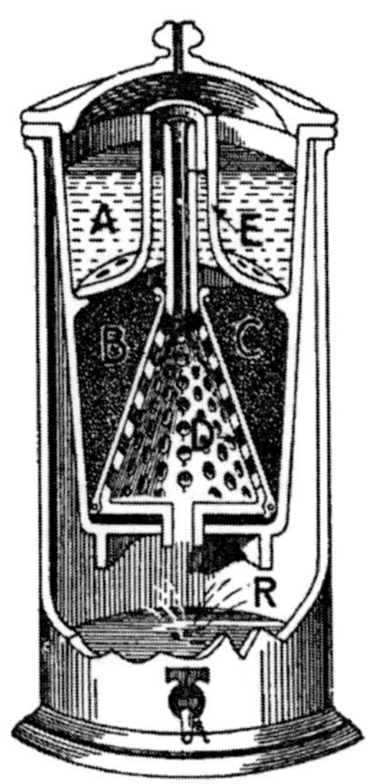

Fig. 185.

The most useful form of household filter is one which can be attached to a tap connected with the main. Such a filter is usually made of porcelain or biscuit china. The Berkefeld filter has an outer case of iron, and an interior hollow "candle" of porcelain from which a tube passes through the lid of the filter to a storage tank for the filtered water. The water from the main enters the outer case, and percolates through the porcelain walls to the internal cavity and thence flows away through the delivery pipe.

Whatever be the type of filter used it must be cleansed at proper intervals. A foul filter is very dangerous to those who drink the water from it. It has been proved by tests that, so far from purifying the water, an inefficient and contaminated filter passes out water much more highly charged with microbes than it was before it entered. We must not therefore think that, because water has been filtered, it is necessarily safe. The reverse is only too often the case.

GAS TRAPS.

Dangerous microbes can be breathed as well as drunk into the human system. Every communication between house and drains should be most carefully "trapped." The principle of a gas trap between, say, a kitchen sink and the drain to carry off the water is given in Fig. 186. Enough water always remains in the bend to rise above the level of the elbow, effectually keeping back any gas that there may be in the pipe beyond the bend.

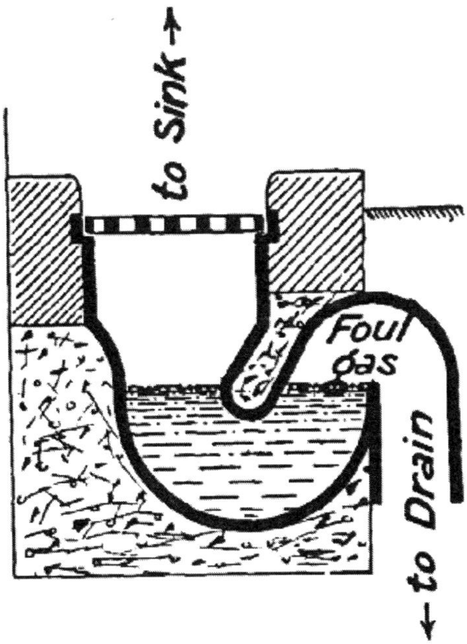

Fig. 186.—A trap for foul air.

WATER-ENGINES.

Before the invention of the steam-engine human industries were largely dependent on the motive power of the wind and running water. But when the infant nursed by Watt and Stephenson had grown into a giant, both of these natural agents were deposed from the important position they once held. Windmills in a state of decay crown many of our hilltops, and the water-wheel which formerly brought wealth to the miller now rots in its mountings at the end of the dam. Except for pumping and moving boats and ships, wind-power finds its occupation gone. It is too uncertain in quantity and quality to find a place in modern economics. Water-power, on the other hand, has received a fresh lease of life through the invention of machinery so scientifically designed as to use much more of the water's energy than was possible with the old-fashioned wheel.

Fig. 187.—A Pelton wheel which develops 5,000 horse-power. Observe the shape of the double buckets.

The *turbine,* of which we have already spoken in our third chapter, is now the favourite hydraulic engine. Some water-turbines work on much the same principle as the Parsons steam-turbine; others resemble the De Laval. Among the latter the Pelton wheel takes the first place. By the courtesy of the manufacturers we are able to give some interesting details and illustrations of this device.

Fig. 188.—Pelton wheel mounted, with nozzle in position.

The wheel, which may be of any diameter from six inches to ten feet, has buckets set at regular intervals round the circumference, sticking outwards. Each bucket, as will be gathered from our illustration of an enormous 5,000 h.p. wheel (Fig. 187), is composed of two cups. A nozzle is so arranged as to direct water on the buckets just as they reach the lowest point of a revolution (see Fig. 188). The water strikes the bucket on the partition between the two cups, which turns it right and left round the inside of the cups. The change of direction transfers the energy of the water to the wheel.

Fig. 189.—Speed regulator for Pelton wheel.

The speed of the wheel may be automatically regulated by a deflecting nozzle (Fig. 189), which has a ball and socket joint to permit of its being raised or lowered by a centrifugal governor, thus throwing the stream on or off the buckets. The power of the wheel is consequently increased or diminished to meet the change of load, and a constant speed is maintained. When it is necessary to waste as little water as possible, a concentric tapered needle may be fitted inside the nozzle. When the nozzle is in its highest position the needle tip is withdrawn; as the nozzle sinks the needle protrudes, gradually decreasing the discharge area of the nozzle.

Pelton wheels are designed to run at all speeds and to use water of any pressure. At Manitou, Colorado, is an installation of three wheels operated by water which leaves the nozzle at the enormous pressure of 935 lbs. per square inch. It is interesting to note that jets of very high-pressure water offer astonishing resistance to any attempt to deflect their course. A three-inch jet of 500-lb. water cannot be cut through by a blow from a crowbar.

In order to get sufficient pressure for working hydraulic machinery in mines, factories, etc., water is often led for many miles in flumes, or artificial channels, along the sides of valleys from the source of supply to the point at which it is to be used. By the time that point is reached the difference between the gradients of the flume and of the valley bottom has produced a difference in height of some hundreds of feet.

Fig. 190.—The Laxey water-wheel, Isle of Man. In the top right-hand corner is a Pelton wheel of proportionate size required to do the same amount of work with the same consumption of water at the same pressure.

The full-page illustration affords a striking testimony to the wonderful progress made in engineering practice during the last fifty years. The huge water-wheel which forms the bulk of the picture is that at Laxey, in the Isle of Man. It is 72½ feet in diameter, and is supposed to develop 150 horse-power, which is transmitted several hundreds of feet by means of wooden rods supported at regular intervals. The power thus transmitted operates a system of pumps in a lead mine, raising 250 gallons of water per minute, to an elevation of 1,200 feet. The driving water is brought some distance to the wheel in an underground conduit, and is carried up the masonry tower by pressure, flowing over the top into the buckets on the circumference of the wheel.

The little cut in the upper corner represents a Pelton wheel drawn on the same scale, which, given an equal supply of water at the same pressure, would develop the same power as the Laxey monster. By the side of the giant the other appears a mere toy.

THE CREAM SEPARATOR.

In 1864 Denmark went to war with Germany, and emerged from the short struggle shorn of the provinces of Lauenburg, Holstein, and Schleswig. The loss of the two last, the fairest and most fertile districts of the kingdom, was indeed grievous. The Danish king now ruled only over a land consisting largely of moor, marsh, and dunes, apparently worthless for any purpose. But the Danes, with admirable courage, entered upon a second struggle, this time with nature. They made roads and railways, dug irrigation ditches, and planted forest trees; and so gradually turned large tracts of what had been useless country into valuable possessions. Agriculture being much depressed, owing to the low price of corn, they next gave their attention to the improvement of dairy farming. Labour-saving machinery of all kinds was introduced, none more important than the device for separating the fatty from the watery constituents of milk. It would not be too much to say that the separator is largely responsible for the present prosperity of Denmark.

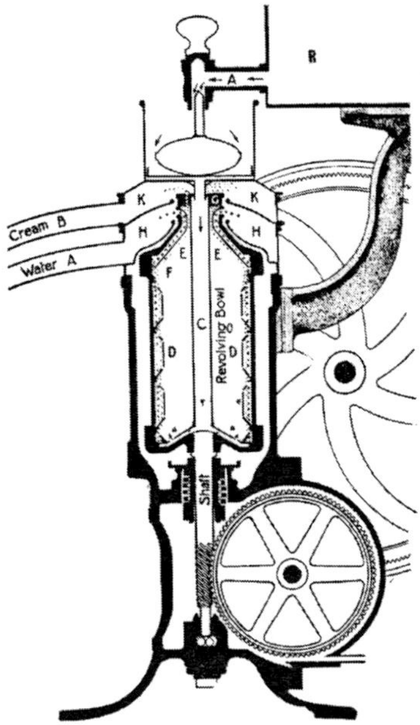

Fig. 191.—Section of a Cream Separator.

How does it work? asks the reader. Centrifugal force[37] is the governing principle. To explain its application we append a sectional illustration (Fig. 191) of Messrs. Burmeister and Wain's hand-power separator, which may be taken as generally representative of this class of machines. Inside a circular casing is a cylindrical bowl,D, mounted on a shaft which can be revolved 5,000 times a minute by means of the cog-wheels and the screw thread chased on it near the bottom extremity. Milk flows from the reservoirR(supported on a stout arm) through tapAinto a little distributer on the top of the separator, and from it drops into the central tubeCof the bowl. Falling to the bottom, it is flung outwards by centrifugal force, finds an escape upwards through the holes *a a*, and climbs up the perforated grid *e*, the surface of which is a series of pyramidical excrescences, and finally reaches the inner surface of the drum proper. The velocity of rotation is so tremendous that the heavier portions of the milk—that is, the watery—crowd towards the point furthest from the centre, and keep the lighter fatty elements away from contact with the sides of the drum. In the diagram the water is represented by small circles, the cream by small crosses.

As more milk enters the drum it forces upwards what is already there. The cap of the drum has an inner jacket,F, which at the bottom *all but touches* the side of the drum. The distance between them is the merest slit; but the cream is deflected up outsideFinto spaceE, and escapes through a hole one-sixteenth of an inch in diameter perforating the plateG. The cream is flung into spaceKand trickles out of spoutB, while the water flies into spaceHand trickles away through spoutA.

THE "HYDRO.,"

used in laundries for wringing clothes by centrifugal force, has a solid outer casing and an inner perforated cylindrical cage, revolved at high speed by a vertical shaft. The wet clothes are placed in the cage, and the machine is started. The water escapes through the perforations and runs down the side of the casing to a drain. After a few minutes the clothes are dry enough for ironing. So great is the centrifugal force that they are consolidated against the sides of the cage, and care is needed in their removal.

[35] Inventor of the lathe slide-rest.

[36] Living germs; some varieties the cause of disease.

[37] That is, centre-fleeing force. Water dropped on a spinning top rushes towards the circumference and is shot off at right angles to a line drawn from the point of parting to the centre of the top.

CHAPTER XIX.
HEATING AND LIGHTING.

The hot-water supply—The tank system—The cylinder system—How a lamp works—Gas and gasworks—Automatic stoking—A gas governor—The gas meter—Incandescent gas lighting.

HOT-WATER SUPPLY.

AWELL-EQUIPPED house is nowadays expected to contain efficient apparatus for supplying plenty of hot water at all hours of the day. There is little romance about the kitchen boiler and the pipes which the plumber and his satellites have sometimes to inspect and put right, but the methods of securing a proper circulation of hot water through the house are sufficiently important and interesting to be noticed in these pages.

In houses of moderate size the kitchen range does the heating. The two systems of storing and distributing the heated water most commonly used are—(1) The *tank* system; (2) the *cylinder* system.

THE TANK SYSTEM

is shown diagrammatically in Fig. 192. The boiler is situated at the back of the range, and when a "damper" is drawn the fire and hot gases pass under it to a flue leading to the chimney. The almost boiling water rises to the top of the boiler and thence finds its way up the *flow pipe* into the hot-water tankA, displacing the somewhat colder water there, which descends through the *return pipe* to the bottom of the boiler.

Water is drawn off from the flow pipe. This pipe projects some distance through the bottom ofA, so that the hottest portion of the contents may be drawn off first. A tank situated in the roof, and fed from the main by a ball-cock valve, communicates withAthrough the siphon pipeS. The bend in this pipe prevents the ascent of hot water, which cannot sink through water colder than itself. From the top ofAan *expansion pipe* is led up and turned over the cold-water tank to discharge any steam which may be generated in the boiler.

A hot-water radiator for warming the house may be connected to the flow and return pipes as shown. Since it opens a "short circuit" for the circulation, the water in the tank above will not be so well heated while it is in action. If cocks are fitted to the radiator pipes, the amount of heat thus deflected can be governed.

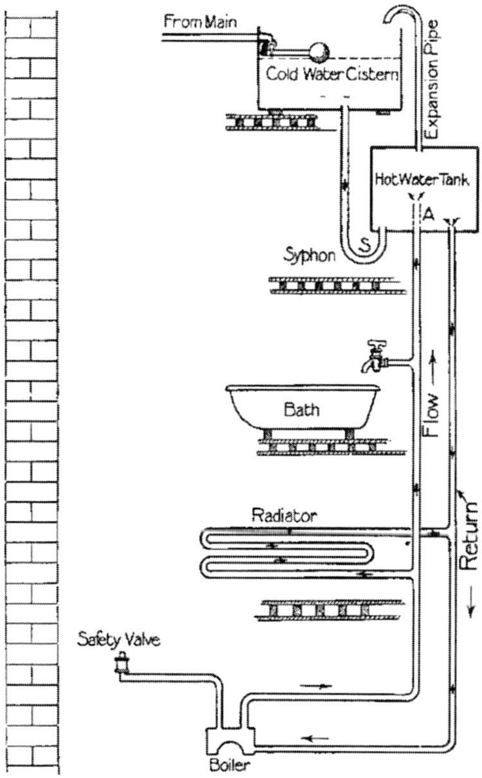

Fig. 192.—The "tank" system of hot-water supply.

A disadvantage of the tank system is that the tank, if placed high enough to supply all flows, is sometimes so far from the boiler that the water loses much of its heat in the course of circulation. Also, if for any reason the cold water fails, tankAmay be entirely emptied, circulation cease, and the water in the boiler and pipes boil away rapidly.

THE CYLINDER SYSTEM

(Fig. 193) is open to neither of these objections. Instead of a rectangular tank up aloft, we now have a large copper cylinder situated in the kitchen near the range. The flow and return pipes are continuous, and the cold supply enters the bottom of the cylinder through a pipe with a siphon bend in it.

As before, water is drawn off from the flow pipe, and a radiator may be put in the circuit. Since there is no draw-off point below the top of the cylinder, even if the cold supply fails the cylinder will remain full, and the failure will be discovered long before there is any danger of the water in it boiling away.

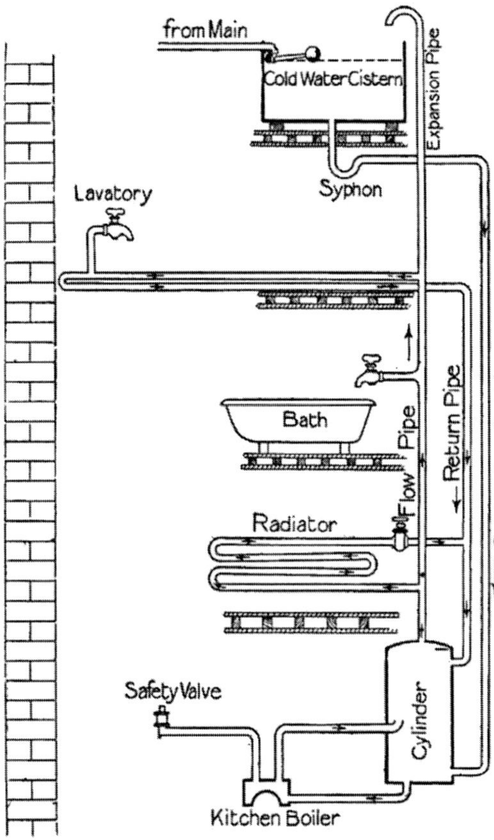

Fig. 193.—The "cylinder" system of hot-water supply.

Boiler explosions are due to obstructions in the pipes. If the expansion pipe and the cold-water supply pipe freeze, there is danger of a slight accumulation of steam; and if one of the circulation pipes is also blocked, steam must generate until "something has to go,"[38] which is naturally the boiler. Assuming that the pipes are quite full to the points of obstruction, the fracture would result from the expansion of the water. Steam cannot generate unless there be a space above the water. But the expanding water has stored up the heat which would have raised steam, and the moment expansion begins after fracture this energy is suddenly let loose. Steam forms instantaneously, augmenting the effects of the explosion. From this it will be gathered that all pipes should be properly protected against frost; especially near the roof.

Another cause of disaster is the *furring up* of the pipes with the lime deposited by hard water when heated. When hard water is used, the pipes will sooner or later be blocked near the boiler; and as the deposit is too hard to be scraped away, periodical renewals are unavoidable.

HOW A LAMP WORKS.

From heating we turn to lighting, and first to the ordinary paraffin lamp. The two chief things to notice about this are the wick and the chimney. The wick, being made of closely-woven cotton, draws up the oil by what is known as *capillary attraction*. If you dip the ends of two glass tubes, one half an inch, the other one-eighth of an inch in diameter, into a vessel of water, you will notice that the water rises higher in the smaller tube. Or get two clean glass plates and lay them face to face, touching at one end, but kept slightly apart at the other by some small object. If they are partly submerged perpendicularly, the water will rise between the plates—furthest on the side at which the two plates touch, and less and less as the other edge is approached. The tendency of liquids to rise through porous bodies is a phenomenon for which we cannot account.

Mineral oil contains a large proportion of carbon and hydrogen; it is therefore termed hydro-carbon. When oil reaches the top of a lighted wick, the liquid is heated until it turns into gas. The carbon and hydrogen unite with the oxygen of the air. Some particles of the carbon apparently do not combine at once, and as they pass through the fiery zone of the flame are heated to such a temperature as to become highly luminous. It is to produce these light-rays that we use a lamp, and to burn our oil efficiently we must supply the flame with plenty of oxygen, with more than it could naturally obtain. So we surround it with a transparent chimney of special glass. The air inside the chimney is heated, and rises; fresh air rushes in at the bottom, and is also heated and replaced. As the air passes through, the flame seizes on the oxygen. If the wick is turned up until the flame becomes smoky and flares, the point has been passed at which the induced chimney draught can supply sufficient oxygen to combine with the carbon of the vapour, and the "free" carbon escapes as smoke.

The blower-plate used to draw up a fire (Fig. 194) performs exactly the same function as the lamp chimney, but on a larger scale. The plate prevents air passing straight up the chimney over the coals, and compels it to find a way through the fire itself to replace the heated air rising up the chimney.

Fig. 194.—Showing how a blower-plate draws up the fire.

GAS AND GASWORKS.

A lamp is an apparatus for converting hydro-carbon mineral oil into gas and burning it efficiently. The gas-jet burns gases produced by driving off hydro-carbon vapours from coal in apparatus specially designed for the purpose. Gas-making is now, in spite of the competition of electric lighting, so important an industry that we shall do well to glance at the processes which it includes. Coal gas may be produced on a very small scale as follows:—Fill a tin canister (the joints of which have been made by folding the metal, not by soldering) with coal, clap on the lid, and place it, lid downwards, in a bright fire, after punching a hole in the bottom. Vapour soon begins to issue from the hole. This is probably at first only steam, due to the coal being more or less damp. But if a lighted match be presently applied the vapour takes fire, showing that coal gas proper is coming off. The flame lasts for a long time. When it dies the canister may be removed and the contents examined. Most of the carbon remains in the form of *coke*. It is bulk for bulk much lighter than coal, for the hydrogen, oxygen, and other gases, and some of the carbon have been driven off by the heat. The coke itself burns if placed in a fire, but without any smoke, such as issues from coal.

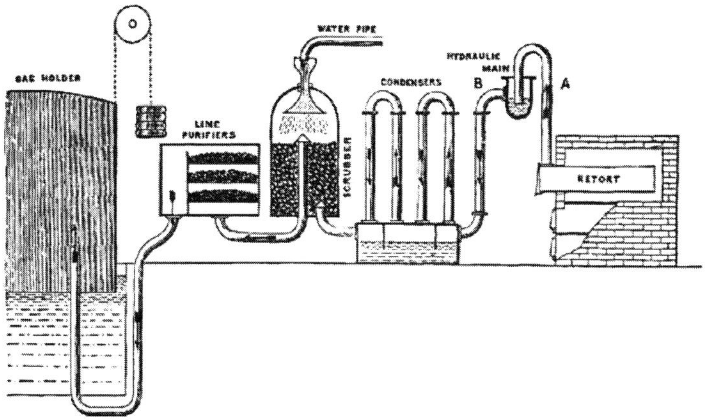

Fig. 195.—Sketch of the apparatus used in the manufacture of coal gas.

Our home-made gas yields a smoky and unsatisfactory flame, owing to the presence of certain impurities—ammonia, tar, sulphuretted hydrogen, and carbon bisulphide. A gas factory must be equipped with means of gettıng rid of these objectionable constituents. Turning to Fig. 195, which displays very diagrammatically the main features of a gas plant, we observe at the extreme right the *retorts*, which correspond to our canister. These are usually long fire-brick tubes ofD-section, the flat side at the bottom. Under each is a furnace, the flames of which play on the bottom, sides, and inner end of the retort. The outer end projecting beyond the brickwork seating has an iron air-tight door for filling the retort through, immediately behind which rises an iron exit pipe,A, for the gases. Tar, which vaporizes at high temperatures, but liquefies at ordinary atmospheric heat, must first be got rid of. This is effected by passing the gas through the *hydraulic main*, a tubular vessel half full of water running the whole length of the retorts. The end of pipeAdips below the surface of the water, which condenses most of the tar and steam. The partly-purified gas now passes through pipeBto the *condensers*, a series of invertedU-pipes standing on an iron chest with vertical cross divisions between the mouths of eachU. These divisions dip into water, so that the gas has to pass up one leg of a U, down the other, up the first leg of the second pipe, and so on, till all traces of the tar and other liquid constituents have condensed on the inside of the pipe, from which they drop into the tank below.

The next stage is the passage of the *scrubber*, filled with coke over which water perpetually flows. The ammonia gas is here absorbed. There still remain the sulphuretted hydrogen and the carbon bisulphide, both of which are extremely offensive to the nostrils. Slaked lime, laid on trays in an air-tight compartment called the *lime purifier*, absorbs most of the sulphurous elements of these; and the coal gas is then fit for use. On leaving the purifiers it flows into the *gasometer*, or gasholder, the huge cake-like form of which is a very familiar object in the environs of towns. The gasometer is a cylindrical box with a domed top, but no bottom, built of riveted steel plates. It stands in a circular tank of water, so that it may rise and fall without any escape of gas. The levity of the gas, in conjunction with weights attached to the ends of chains working over pulleys on the framework surrounding the holder, suffices to raise the holder.

Fig. 196.—The largest gasholder in the world: South Metropolitan Gas Co., Greenwich Gas Works. Capacity, 12,158,600 cubic feet.

Some gasometers have an enormous capacity. The record is at present held by that built for the South Metropolitan Gas Co., London, by Messrs. Clayton & Son of Leeds. This monster (of which we append an illustration, Fig. 196) is 300 feet in diameter and 180 feet high. When fully extended it holds 12,158,600 cubic feet of gas. Owing to its immense size, it is built on the telescopic principle in six "lifts," of 30 feet deep each. The sides of each lift, or ring, except the topmost, have a section shaped somewhat like the letter N. Two of the members form a deep, narrow cup to hold water, in which the "dip" member of the ring above it rises and falls.

Fig. 197.—Drawing retorts. (*Photo by F. Marsh.*)

AUTOMATIC STOKING.

The labour of feeding the retorts with coal and removing the coke is exceedingly severe. In the illustration (made from a very fine photograph taken by Mr. F. Marsh of Clifton) we see a man engaged in "drawing" the retorts through the iron doors at their outer ends. Automatic machinery is now used in large gasworks for both operations. One of the most ingenious stokers is the De Brouwer, shown at work in Fig. 198. The machine is suspended from an overhead trolley running on rails along the face of the retorts. Coal falls into a funnel at the top of the telescopic pipePfrom hoppers in the story above, which have openings,H H, controlled by shutters. The coal as it falls is caught by a rubber belt working round part of the circumference of the large wheelWand a number of pulleys, and is shot into the mouth of the retort. The operator is seen pulling the handle which opens the shutter of the hopper above the feed-tube, and switching on the 4 h.p. electric motor which drives the belt and moves the machine about. One of these feeders will charge a retort 20 feet long in twenty-two seconds.

Fig. 198.—De Brouwer automatic retort charger.

A GAS GOVERNOR.

Some readers may have noticed that late at night a gas-jet, which a few hours before burned with a somewhat feeble flame when the tap was turned fully on, now becomes more and more vigorous, and finally may flare up with a hissing sound. This is because many of the burners fed by the main supplying the house have been turned off, and consequently there is a greater amount of gas available for the jets still burning, which therefore feel an increased pressure. As a matter of fact, the pressure of gas in the main is constantly varying, owing partly to the irregularity of the delivery from the gasometer, and partly to the fact that the number of burners in action is not the same for many minutes together. It must also be remembered that houses near the gasometer end of the main will receive their gas at a higher pressure than those at the other end. The gas stored in the holders may be wanted for use in the street lamps a few yards away, or for other lamps several miles distant. It is therefore evident that if there be just enough pressure to give a good supply to the nearest lamp, there will be too little a short distance beyond it, and none at all at the extreme point; so that it is necessary to put on enough pressure to overcome the friction on all these miles of pipe, and give just enough gas at the extreme end. It follows that at all intermediate points the pressure is excessive. Gas of the average quality is burned to the greatest advantage, as regards its light-giving properties, when its pressure is equal to that of a column of water half an inch high, or about1/50 lb. to the square inch. With less it gives a smoky, flickering light, and with more the combustion is also imperfect.

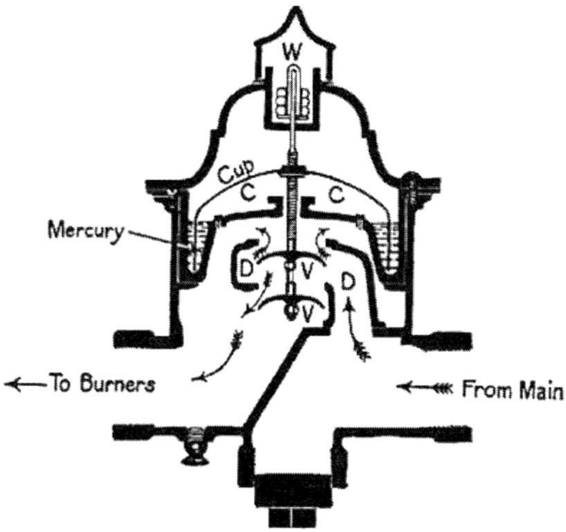

Fig. 199.

Every house supply should therefore be fitted with a gas governor, to keep the pressure constant. A governor frequently used, the Stott, is shown in section in Fig. 199. Gas enters from the main on the right, and passes into a circular elbow,D, which has top and bottom apertures closed by the valvesV V. Attached to the valve shaft is a large inverted cup of metal, the tip of which is immersed in mercury. The pressure at which the governor is to act is determined by the weightsW, with which the valve spindle is loaded at the top. As soon as this pressure is exceeded, the gas inC Clifts the metal cup, andV Vare pressed against their seats, so cutting off the supply. Gas cannot escape fromC C, as it has not sufficient pressure to force its way through the mercury under the lip of the cup. Immediately the pressure inC Cfalls, owing to some of the gas being used up, the valves open and admit more gas. When the fluctuations of pressure are slight, the valves never close completely, but merely throttle the supply until the pressure beyond them falls to its proper level—that is, they pass just as much gas as the burners in use can consume at the pressure arranged for.

Governors of much larger size, but working on much the same principle, are fitted to the mains at the point where they leave the gasometers. They are not, however, sensitive to local fluctuations in the pipes, hence the necessity for separate governors in the house between the meter and the burners.

THE GAS-METER

commonly used in houses acts on the principle shown in Fig. 200. The air-tight casing is divided by horizontal and vertical divisions into three gas-chambers,B,C, andD. Gas enters atA, and passes to the valve chamberB. The slide-valves of this allow it to pass intoCandD, and also into the two circular leather bellowsE, F, which are attached to the central divisionG, but are quite independent of one another.

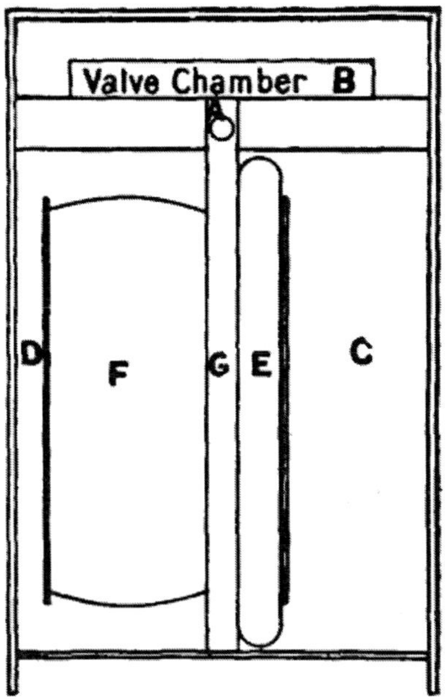

Fig. 200.—Sketch of the bellows and chambers of a "dry" gas meter.

We will suppose that in the illustration the valves are admitting gas to chamberCand bellowsF. The pressure inCpresses the circular head ofEtowards the divisionG, expelling the contents of the bellows through an outlet pipe (not shown) to the burners in operation within the house. Simultaneously the inflation ofFforces the gas in chamberDalso through the outlet. The head-plates of the bellows are attached to rods and levers (not shown) working the slide-valves inB. As soon asEis fully in, andFfully expanded, the valves begin to open and put the inlet pipe in communication withDandE, and allow the contents ofFandCto escape to the outlet. The movements of the valve mechanism operate a train of counting wheels, visible through a glass window in the side of the case. As the bellows have a definite capacity, every stroke that they give means that a certain volume of gas has been ejected either from them or from the chambers in which they move: this is registered by the counter. The apparatus practically has two double-action cylinders (of which the bellows ends are the pistons) working on the same principle as the steam-cylinder (Fig. 21). The valves have three ports—the central, or exhaust, leading to the outlet, the outer ones from the inlet. The bellows are fed through channels in the divisionG.

INCANDESCENT GAS LIGHTING.

The introduction of the electric arc lamp and the incandescent glow-lamp seemed at one time to spell the doom of gas as an illuminating agent. But the appearance in 1886 of the Welsbach *incandescent mantle* for gas-burners opened a prosperous era in the history of gas lighting.

The luminosity of a gas flame depends on the number of carbon particles liberated within it, and the temperature to which these particles can be heated as they pass through the intensely hot outside zone of the flame. By enriching the gas in carbon more light is yielded, up to a certain point, with a flame of a given temperature. To increase the heat of the flame various devices were tried before the introduction of the incandescent mantle, but they were found to be too short-lived to have any commercial value. Inventors therefore sought for methods by which the emission of light could be obtained from coal gas independently of the incandescence of the carbon particles in the flame itself; and step by step it was discovered that gas could be better employed merely as a heating agent, to raise to incandescence substances having a higher emissivity of light than carbon.

Dr. Auer von Welsbach found that the substances most suitable for incandescent mantles were the oxides of certain rare metals, *thorium*, and *cerium*. The mantle is made by dipping a cylinder of cotton net into a solution of nitrate of thorium and cerium, containing 99 per cent. of the former and 1 per cent. of the latter metal. When the fibres are sufficiently soaked, the mantle is withdrawn, squeezed, and placed on a mould to dry. It is next held over a Bunsen gas flame and the cotton is burned away, while the nitrates are converted into oxides. The mantle is now ready for use, but very brittle. So it has to undergo a further dipping, in a solution of gun-cotton and alcohol, to render it tough enough for packing. When it is required for use, it is suspended over the burner by an asbestos thread woven across the top, a light is applied to the bottom, and the collodion burned off, leaving nothing but the heat-resisting oxides.

The burner used with a mantle is constructed on the Bunsen principle. The gas is mixed, as it emerges from the jet, with sufficient air to render its combustion perfect. All the carbon is burned, and the flame, though almost invisible, is intensely hot. The mantle oxides convert the heat energy of the flame into light energy. This is proved not only by the intense whiteness of the mantle, but by the fact that the heat issuing from the chimney of the burner is not nearly so great when the mantle is in position as when it is absent.

The incandescent mantle is more extensively used every year. In Germany 90 per cent. of gas lighting is on the incandescent system, and in England about 40 per cent. We may notice, as an interesting example of the fluctuating fortunes of invention, that the once doomed gas-burner has, thanks to Welsbach's mantle, in many instances replaced the incandescent electric lamps that were to doom it.

[38] If, of course, there is no safety-valve in proper working order included in the installation.

CHAPTER XX.
VARIOUS MECHANISMS.

Clocks and Watches:—A short history of timepieces—The construction of timepieces—The driving power—The escapement—Compensating pendulums—The spring balance—The cylinder escapement—The lever escapement—Compensated balance-wheels—Keyless winding mechanism for watches—The hour hand train.Locks:—The Chubb lock—The Yale lock.The Cycle:—The gearing of a cycle—The free wheel—The change-speed gear.Agricultural Machines:—The threshing-machine—Mowing-machines.Some Natural Phenomena:—Why sun-heat varies in intensity—The tides—Why high tide varies daily.

CLOCKS AND WATCHES.

A SHORT HISTORY OF TIMEPIECES.

THE oldest device for measuring time is the sun-dial. That of Ahaz mentioned in the Second Book of Kings is the earliest dial of which we have record. The obelisks of the Egyptians and the curious stone pillars of the Druidic age also probably served as shadow-casters.

The clepsydra, or water-clock, also of great antiquity, was the first contrivance for gauging the passage of the hours independently of the motion of the earth. In its simplest form it was a measure into which water fell drop by drop, hour levels being marked on the inside. Subsequently a very simple mechanism was added to drive a pointer—a float carrying a vertical rack, engaging with a cog on the pointer spindle; or a string from the float passed over a pulley attached to the pointer and rotated it as the float rose, after the manner of the wheel barometer (Fig. 153). In 807A.D.Charlemagne received from the King of Persia a water-clock which struck the hours. It is thus described in Gifford's "History of France":—"The dial was composed of twelve small doors, which represented the division of the hours. Each door opened at the hour it was intended to represent, and out of it came a small number of little balls, which fell one by one, at equal distances of time, on a brass drum. It might be told by the eye what hour it was by the number of doors that were open, and by the ear by the number of balls that fell. When it was twelve o'clock twelve horsemen in miniature issued forth at the same time and shut all the doors."

Sand-glasses were introduced about 330A.D.Except for special purposes, such as timing sermons and boiling eggs, they have not been of any practical value.

The clepsydra naturally suggested to the mechanical mind the idea of driving a mechanism for registering time by the force of gravity acting on some body other than water. The invention of the *weight-driven clock* is attributed, like a good many other things, to Archimedes, the famous Sicilian mathematician of the third centuryB.C.; but no record exists of any actual clock composed of wheels operated by a weight prior to 1120A.D.So we may take that year as opening the era of the clock as we know it.

About 1500 Peter Hele of Nuremberg invented the *mainspring* as a substitute for the weight, and the *watch* appeared soon afterwards (1525A.D.). The pendulum was first adopted for controlling the motion of the wheels by Christian Huygens, a distinguished Dutch mechanician, in 1659.

To Thomas Tompion, "the father of English watchmaking," is ascribed the honour of first fitting a *hairspring* to the escapement of a watch, in or about the year 1660. He also introduced the *cylinder escapement* now so commonly used in cheap watches. Though many improvements have been made since his time, Tompion manufactured clocks and watches which were excellent timekeepers, and as a reward for the benefits conferred on his fellows during his lifetime, he was, after death, granted the exceptional honour of a resting-place in Westminster Abbey.

THE CONSTRUCTION OF TIMEPIECES.

A clock or watch contains three main elements:—(1) The source of power, which may be a weight or a spring; (2) the train of wheels operated by the driving force; (3) the agent for controlling the movements of the train—this in large clocks is usually a pendulum, in small clocks and watches a hairspring balance. To these may be added, in the case of clocks, the apparatus for striking the hour.

THE DRIVING POWER.

Weights are used only in large clocks, such as one finds in halls, towers, and observatories. The great advantage of employing weights is that a constant driving power is exerted. *Springs* occupy much less room than weights, and are indispensable for portable timepieces. The employment of them caused trouble to early experimenters on account of the decrease in power which necessarily accompanies the uncoiling of a wound-up spring. Jacob Zech of Prague overcame the difficulty in 1525 by the invention of the *fusee,* a kind of conical pulley interposed between the barrel, or circular drum containing the mainspring, and the train of wheels which the spring has to drive. The principle of the "drum and fusee" action will be understood from Fig. 201. The mainspring is a long steel ribbon fixed at one end to an arbor (the watchmaker's name for a spindle or axle), round which it is tightly wound. The arbor and spring are inserted in the barrel. The arbor is prevented from turning by a ratchet,B, and click, and therefore the spring in its effort to uncoil causes the barrel to rotate.

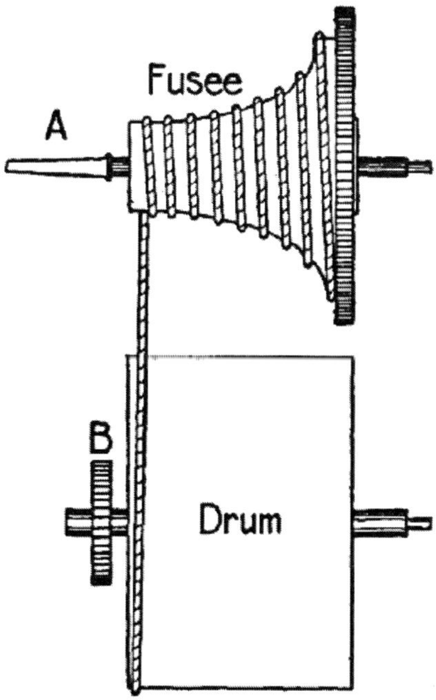

Fig. 201.

A string of catgut (or a very fine chain) is connected at one end to the circumference of the drum, and wound round it, the other end being fixed to the larger end of the fusee, which is attached to the driving-wheel of the watch or clock by the intervention of a ratchet and click (not shown). To wind the spring the fusee is turned backward by means of a key applied to the square endAof the fusee arbor, and this draws the string from off the drum on to the fusee. The force of the spring causes the fusee to rotate by pulling the string off it, coil by coil, and so drives the train of wheels. But while the mainspring, when fully wound, turns the fusee by uncoiling the string from the smallest part of the fusee, it gets the advantage of the larger radius as its energy becomes lessened.

The fusee is still used for marine chronometers, for some clocks that have a mainspring and pendulum, and occasionally for watches. In the latter it has been rendered unnecessary by the introduction of the *going-barrel* by Swiss watchmakers, who formed teeth on the edge of the mainspring barrel to drive the train of wheels. This kind of drum is called "going" because it drives the watch during the operation of winding, which is performed by rotating the drum arbor to which the inner end of the spring is attached. A ratchet prevents the arbor from being turned backwards by the spring. The adoption of the going-barrel has been made satisfactory by the improvements in the various escapement actions.

THE ESCAPEMENT.

Fig. 202.

The spring or weight transmits its power through a train of cogs to the *escapement*, or device for regulating the rate at which the wheels are to revolve. In clocks a *pendulum* is generally used as the controlling agent. Galileo, when a student at Pisa, noticed that certain hanging lamps in the cathedral there swung on their cords at an equal rate; and on investigation he discovered the principle that the shorter a pendulum is the more quickly will it swing to and fro. As has already been observed, Huygens first applied the principle to the governing of clocks. In Fig. 202 we have a simple representation of the "dead-beat" escapement commonly used in clocks. The escape-wheel is mounted on the shaft of the last cog of the driving train, the pallet on a spindle from which depends a split arm embracing the rod and the pendulum. We must be careful to note that the pendulum *controls* motion only; it does not cause movement.

The escape-wheel revolves in a clockwise direction. The two pallets *a* and *b* are so designed that only one can rest on the teeth at one time. In the sketch the sloping end of *b* has just been forced upwards by the pressure of a tooth. This swings the pallet and the pendulum. The momentum of the latter causes *a* to descend, and at the instant when *b* clears its tooth *a* catches and holds another. The left-hand side of *a*, called the *locking-face*, is part of a circle, so that the escape-wheel is held motionless as long as it touches *a*: hence the term, "dead beat"—that is, brought to a dead stop. As the pendulum swings back, to the left, under the influence of gravity, *a* is raised and frees the tooth. The wheel jerks round, and another tooth is caught by the locking-face of *b*. Again the pendulum swings to the right, and the sloping end of *b* is pushed up once more, giving the pendulum fresh impetus. This process repeats itself as long as the driving power lasts—for weeks, months, or years, as the case may be, and the mechanism continues to be in good working order.

COMPENSATING PENDULUMS.

Metal expands when heated; therefore a steel pendulum which is of the exact length to govern a clock correctly at a temperature of 60° would become too long at 80°, and slow the clock, and too short at 40°, and cause it to gain. In common clocks the pendulum rod is often made of wood, which maintains an almost constant length at all ordinary temperatures. But for very accurate clocks something more efficient is required. Graham, the partner of Thomas Tompion, took advantage of the fact that different kinds of metal have different ratios of expansion to produce a *self-compensating* pendulum on the principle illustrated by Fig. 203. He used steel for the rod, and formed the *bob*, or weighted end, of a glass jar containing mercury held in a stirrup; the mercury being of such a height that, as the pendulum rod lengthened with a rise of temperature, the mercury expanded *upwards* sufficiently to keep the distance between the point of suspension and the centre of gravity of the bob always the same. With a fall of temperature the rod shortened, while the mercury sank in the jar. This device has not been improved upon, and is still used in observatories and other places where timekeepers of extreme precision are required. The milled nutSin Fig. 203 is fitted at the end of the pendulum rod to permit the exact adjustment of the pendulum's length.

For watches, chronometers, and small clocks

THE SPRING BALANCE

takes the place of the pendulum. We still have an escape-wheel with teeth of a suitable shape to give impulses to the controlling agent. There are two forms of spring escapement, but as both employ a hairspring and balance-wheel we will glance at these before going further.

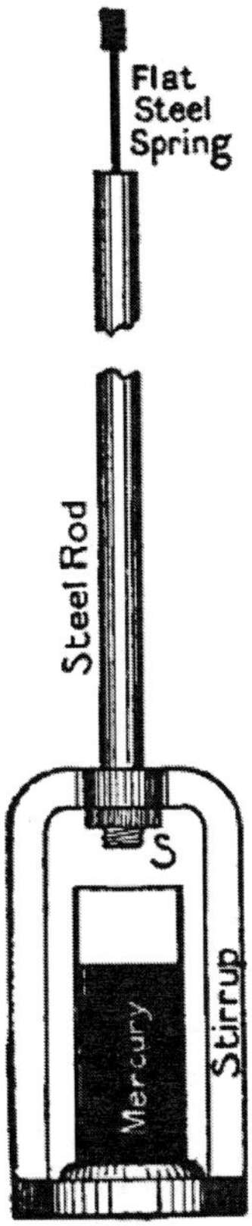

Fig. 203.

The *hairspring* is made of very fine steel ribbon, tempered to extreme elasticity, and shaped to a spiral. The inner end is attached to the arbor of the *balance-wheel*, the outer end to a stud projecting from the plate of the watch. When the balance-wheel, impelled by the escapement, rotates, it winds up the spring. The energy thus stored helps the wheel to revolve the other way during the locking of a tooth of the escape-wheel. The time occupied by the winding and the unwinding depends upon the length of the spring. The strength of the impulse makes no difference. A strong impulse causes the spring to coil itself up more than a weak impulse would; but inasmuch as more energy is stored the process of unwinding is hastened. To put the matter very simply—a strong impulse moves the balance-wheel further, but rotates it quickly; a weak impulse moves it a shorter distance, but rotates it slowly. In fact, the principle of the pendulum is also that of the hairspring; and the duration of a vibration depends on the length of the rod in the one case, and of the spring in the other.

Motion is transmitted to the balance by one of two methods. Either (1) directly, by a cylinder escapement; or (2) indirectly, through a lever.

Fig. 204.—"Cylinder" watch escapement.

THE CYLINDER ESCAPEMENT

is seen in Fig. 204. The escape-wheel has sharp teeth set on stalks. (One tooth is removed to show the stalk.) The balance-wheel is mounted on a small steel cylinder, with part of the circumference cut away at the level of the teeth, so that if seen from above it would appear like *a* in our illustration. A tooth is just beginning to shove its point under the nearer edge of the opening. As it is forced forwards, *b* is revolved in a clockwise direction, winding up the hairspring. When the tooth has passed the nearer edge it flies forward, striking the inside of the further wall of the cylinder, which holds it while the spring uncoils. The tooth now pushes its way past the other edge, accelerating the unwinding, and, as it escapes, the next tooth jumps forward and is arrested by the outside of the cylinder. The balance now reverses its motion, is helped by the tooth, is wound up, locks the tooth, and so on.

THE LEVER ESCAPEMENT

is somewhat more complicated. The escape-wheel teeth are locked and unlocked by the palletsP P1 projecting from a lever which moves on a pivot (Fig. 205). The end of the lever is forked, and has a square notch in it. On the arbor of the balance-wheel is a roller, or plate,R, which carries a small pin,I. Two pins,B B, projecting from the plate of the watch prevent the lever moving too far. We must further notice the little pinCon the lever, and a notch in the edge of the roller.

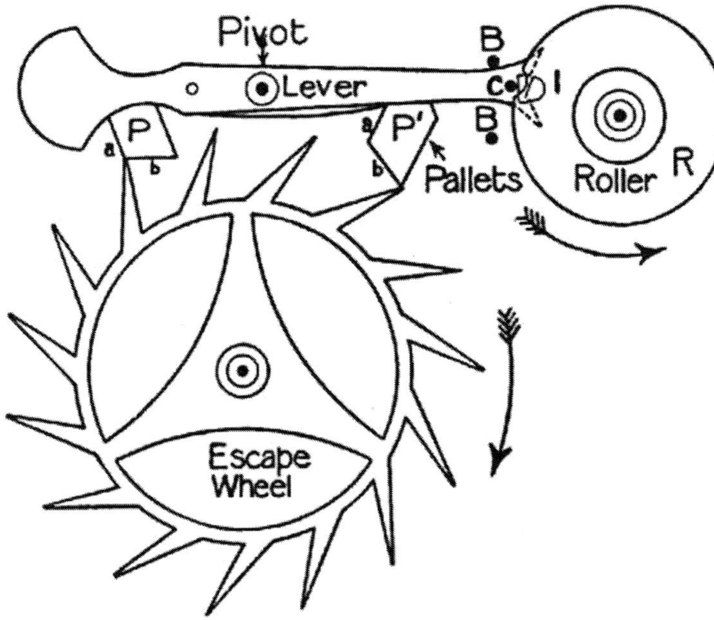

Fig. 205.—"Lever" watch escapement.

In the illustration a tooth has just passed under the "impulse face" *b* ofP1. The lever has been moved upwards at the right end; and its forked end has given an impulse toR, and through it to the balance-wheel. The spring winds up. The pinCprevents the lever dropping, because it no longer has the notch opposite to it, but presses on the circumference ofR. As the spring unwinds it strikes the lever at the moment when the notch andCare opposite. The lever is knocked downwards, and the tooth, which had been arrested by the locking-face *a* of palletP, now presses on the impulse face *b*, forcing the left end of the lever up. The impulse pinIreceives a blow, assisting the unwinding of the spring, andCagain locks the lever. The same thing is repeated in alternate directions over and over again.

COMPENSATING BALANCE-WHEELS.

The watchmaker has had to overcome the same difficulty as the clockmaker with regard to the expansion of the metal in the controlling agent. When a metal wheel is heated its spokes lengthen, and the rim recedes from the centre. Now, let us suppose that we have two rods of equal weight, one three feet long, the other six feet long. To an end of each we fasten a 2-lb. weight. We shall find it much easier to wave the shorter rod backwards and forwards quickly than the other. Why? Because the weight of the longer rod has more leverage over the hand than has that of the shorter rod. Similarly, if, while the mass of the rim of a wheel remains constant, the length of the spokes varies, the effort needed to rotate the wheel to and fro at a constant rate must vary also. Graham got over the difficulty with a rod by means of the compensating pendulum. Thomas Earnshaw mastered it in wheels by means of the *compensating balance*, using the same principle—namely, the unequal expansion of different metals. Any one who owns a compensated watch will see, on stopping the tiny fly-wheel, that it has two spokes (Fig. 206), each carrying an almost complete semicircle of rim attached to it. A close examination shows that the rim is compounded of an outer strip of brass welded to an inner lining of steel. The brass element expands more with heat and contracts more with cold than steel; so that when the spokes become elongated by a rise of temperature, the pieces bend inwards at their free ends (Fig. 207); if the temperature falls, the spokes are shortened, and the rim pieces bend outwards (Fig. 208).[39] This ingenious contrivance keeps the leverage of the rim constant within very fine limits. The screwsS Sare inserted in the rim to balance it correctly, and very fine adjustment is made by means of the four tiny weightsW W. In ships' chronometers,[40] the rim pieces are *sub*-compensated towards their free ends to counteract slight errors in the primary compensation. So delicate is the compensation that a daily loss or gain of only half a second is often the limit of error.

Fig. 206. **Fig. 207.** **Fig. 208.**

A "compensating" watch balance, at normal, super-normal, and sub-normal temperatures.

KEYLESS WINDING MECHANISM FOR WATCHES.

The inconvenience attaching to a key-wound watch caused the Swiss manufacturers to put on the market, in 1851, watches which dispensed with a separate key. Those of our readers who carry keyless watches will be interested to learn how the winding and setting of the hands is effected by the little serrated knob enclosed inside the pendant ring.

There are two forms of "going-barrel" keyless mechanism—(1) The rocking bar; (2) the shifting sleeve. The *rocking bar* device is shown in Figs. 209, 210. The milled headMturns a cog,G, which is always in gear with a cog,F. This cog gears with two others,AandB, mounted at each end of the rockerR, which moves on pivotS. A spring,S P, attached to the watch plate presses against a small stud on the rocking bar, and keepsAnormally in gear withC, mounted on the arbor of the mainspring.

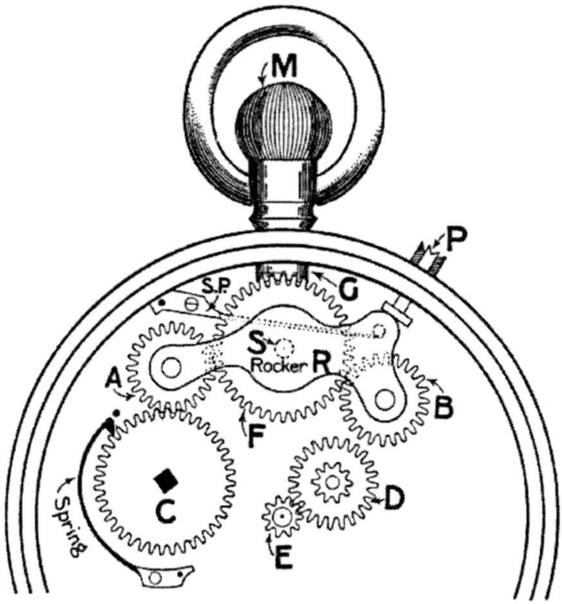

Fig. 209.—The winding mechanism of a keyless watch.

To wind the watch,Mis turned so as to giveFan anti-clockwise motion. The teeth ofFnow pressAdownwards and keep it in gear withCwhile the winding is done. A spring click (marked solid black) prevents the spring uncoiling (Fig. 209). IfFis turned in a clockwise direction it liftsAand prevents it biting the teeth ofC, and no strain is thrown onC.

To set the hands, the little push-piecePis pressed inwards by the thumb (Fig. 210) so as to depress the right-hand end ofRand bringBinto gear withD, which in turn movesE, mounted on the end of the minute-hand shaft. The hands can now be moved in either direction by turningM. On releasing the push-piece the winding-wheels engage again.

The *shifting sleeve* mechanism has a bevel pinion in the place ofG(Fig. 209) gearing with the mainspring cog. The shaft of the knobMis round where it passes through the bevel and can turn freely inside it, but is square below. On the square part is mounted a little sliding clutch with teeth on the top corresponding with the other teeth on the under side of the bevel-wheel, and teeth similar to those ofG(Fig. 209) at the end. The clutch has a groove cut in the circumference, and in this lies the end of a spring lever which can be depressed by the push-piece. The mechanism much resembles on a small scale the motor car changing gear (Fig. 49). Normally, the clutch is pushed up the square part of the knob shaft by the spring so as to engage with the bevel and the winding-wheels. On depressing the clutch by means of the push-piece it gears with the minute-hand pinion, and lets go of the bevel.

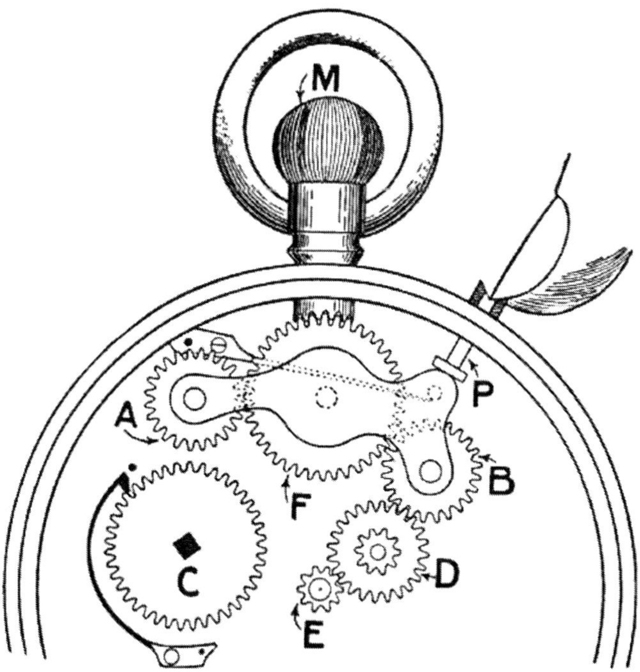

Fig. 210.—The hand-setting mechanism in action.

In one form of this mechanism the push-piece is dispensed with, and the minute-wheel pinion is engaged by pulling the knob upwards.

THE HOUR-HAND TRAIN.

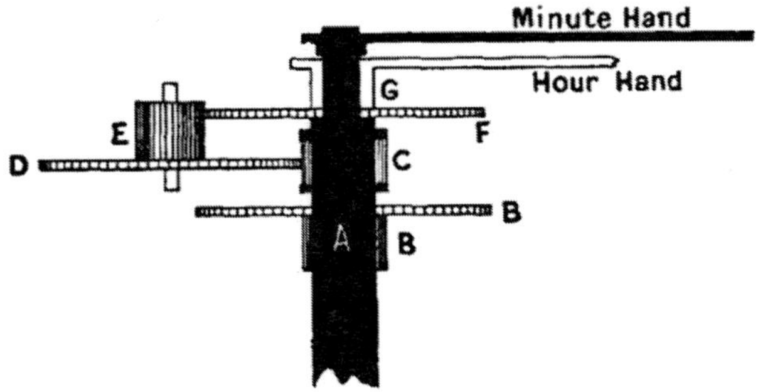

Fig. 211.—The hour-hand train of a clock.

The teeth of the mainspring drum gear with a cog on the minute-hand shaft, which also carries one of the cogs of the escapement train. The shaft is permitted by the escapement to revolve once an hour. Fig. 211 shows diagrammatically how this is managed. The hour-hand shaftA(solid black) can be moved round inside the cogB, driven by the mainspring drum. It carries a cog,C. This gears with a cog,D, having three times as many teeth. The cogE, united toD, drives cogF, having four times as many teeth asE. ToFis attached the collarGof the hour-hand.FandGrevolve outside the minute-hand shaft. On turningA,CturnsDandE,EturnsFand the hour-hand, which revolves ⅓ of ¼ =1⁄12 as fast asA.[41]

LOCKS.

On these unfortunately necessary mechanisms a great deal of ingenuity has been expended. With the advance of luxury and the increased worship of wealth, it becomes more and more necessary to guard one's belongings against the less scrupulous members of society.

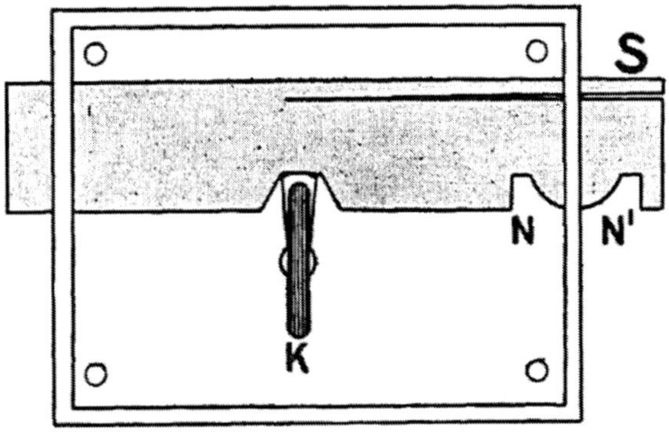

Fig. 212.

The simplest form of lock, such as is found in desks and very cheap articles, works on the principle shown in Fig. 212. The bolt is split at the rear, and the upper part bent upwards to form a spring. The under edge has two notches cut in it, separated by a curved excrescence. The key merely presses the bolt upwards against the spring, until the notch, engaging with the frame, moves it backwards or forwards until the spring drives the tail down into the other notch. This primitive device affords, of course, very little security. An advance is seen in the

TUMBLER LOCK.

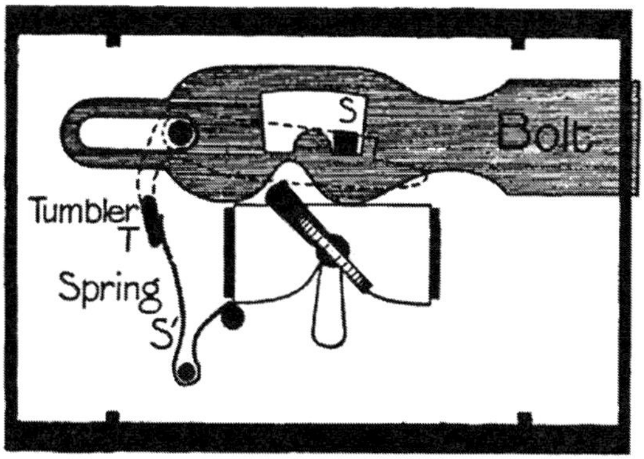

Fig. 213.

The bolt now can move only in a horizontal direction. It has an opening cut in it with two notches (Figs. 213, 214). Behind the bolt lies the *tumbler*T(indicated by the dotted line), pivoted at the angle on a pin. From the face of the tumbler a stud,S, projects through the hole in the bolt. This stud is forced into one or other of the notches by the spring,S1, which presses on the tail of the tumbler.

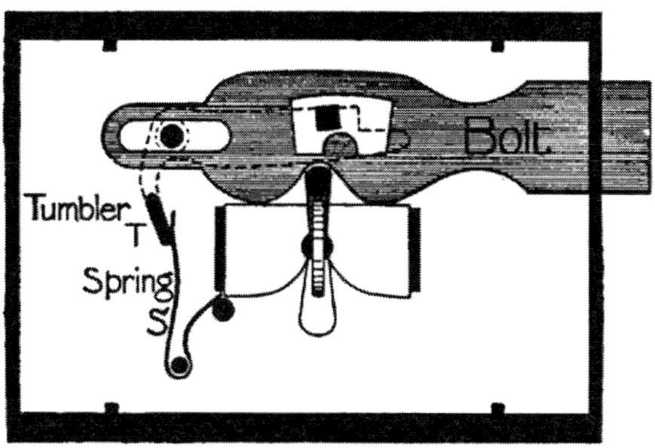

Fig. 214.

In Fig. 213 the key is about to actuate the locking mechanism. The next diagram (Fig. 214) shows how the key, as it enters the notch on the lower side of the bolt to move it along, also raises the tumbler stud clear of the projection between the two notches. By the time that the bolt has been fully "shot," the key leaves the under notch and allows the tumbler stud to fall into the rear locking-notch.

A lock of this type also can be picked very easily, as the picker has merely to lift the tumbler and move the bolt along. Barron's lock, patented in 1778, had two tumblers and two studs; and the opening in the bolt had notches at the top as well as at the bottom (Fig. 215). This made it necessary for both tumblers to be raised simultaneously to exactly the right height. If either was not lifted sufficiently, a stud could not clear its bottom notch; if either rose too far, it engaged an upper notch. The chances therefore were greatly against a wrong key turning the lock.

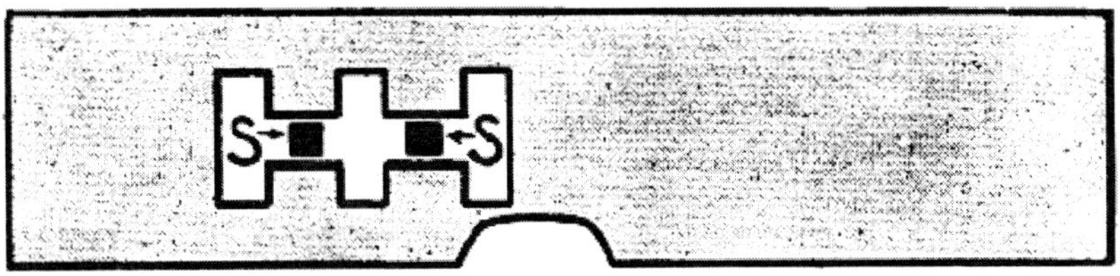

Fig. 215.—The bolt of a Barron lock.

THE CHUBB LOCK

is an amplification of this principle. It usually has several tumblers of the shape shown in Fig. 216. The lock stud in these locks projects from the bolt itself, and the openings, or "gates," through which the stud must pass as the lock moves, are cut in the tumblers. It will be noticed that the forward notch of the tumbler has square serrations in the edges. These engage with similar serrations in the bolt stud and make it impossible to raise the tumbler if the bolt begins to move too soon when a wrong key is inserted.

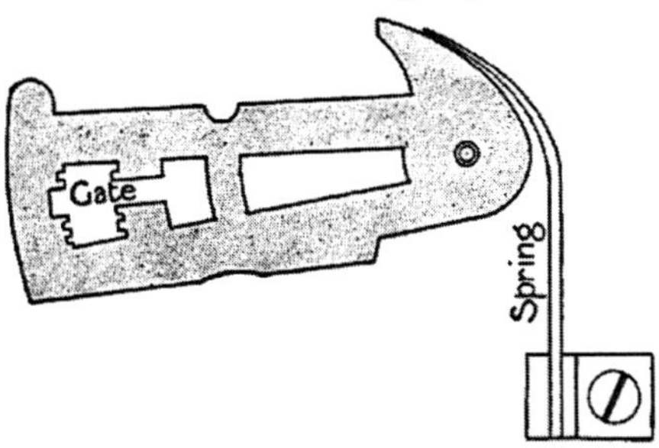

Fig. 216.—Tumbler of Chubb lock.

Fig. 217 is a Chubb key with eight steps. That nearest the head (8) operates a circular revolving curtain, which prevents the introduction of picking tools when a key is inserted and partly turned, as the key slot in the curtain is no longer opposite that in the lock. Step 1 moves the bolt.

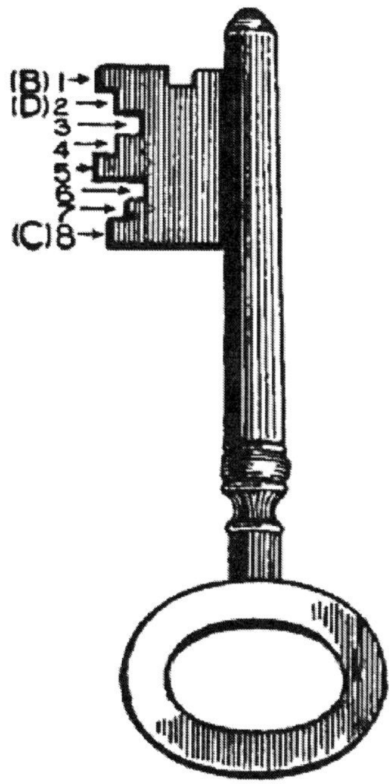

Fig. 217.—A Chubb key.

In order to shoot the bolt the height of the key steps must be so proportioned to the depth of their tumblers that all the gates in the tumblers are simultaneously raised to the right level for the stud to pass through them, as in Fig. 218. Here you will observe that the tumblerDon the extreme right (lifted by step 2 of the key) has a stud,D S, projecting from it over the other tumblers. This is called the *detector tumbler*. If a false key or picking tool is inserted it is certain to raise one of the tumblers too far. The detector is then over-lifted by the studD S, and a spring catch falls into a notch at the rear. It is now impossible to pick the lock, as the detector can be released only by the right key shooting the bolt a little further in the locking direction, when a projection on the rear of the bolt lifts the catch and allows the tumbler to fall. The detector also shows that the lock has been tampered with, since even the right key cannot move the bolt until the overlocking has been performed.

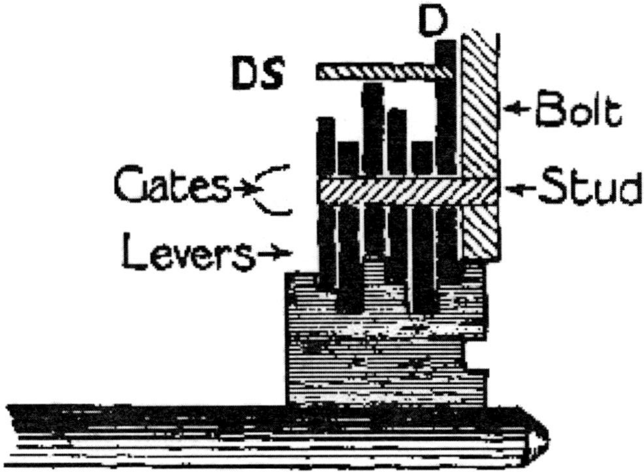

Fig. 218.—A Chubb key raising all the tumblers to the correct height.

Each tumbler step of a large Chubb key can be given one of thirty different heights; the bolt step one of twenty. By merely transposing the order of the steps in a six-step key it is possible to get 720 different combinations. By diminishing or increasing the heights the possible combinations may be raised to the enormous total of 7,776,000!

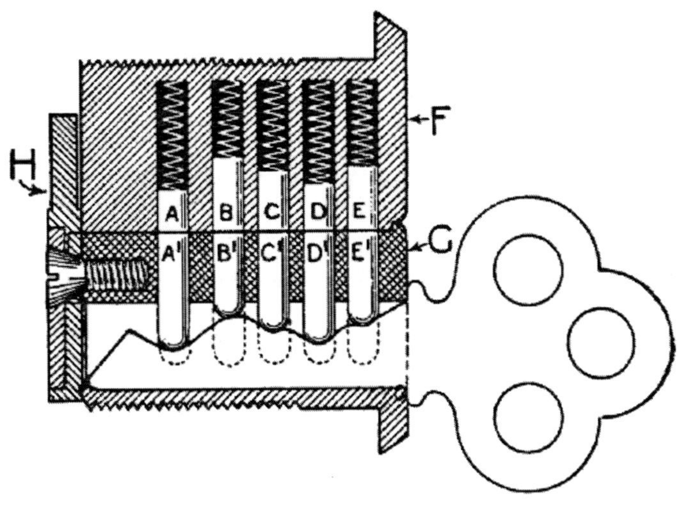

Fig. 219.—Section of a Yale lock.

THE YALE LOCK,

which comes from America, works on a quite different system. Its most noticeable feature is that it permits the use of a very small key, though the number of combinations possible is still enormous (several millions). In our illustrations (Figs. 219, 220, 221) we show the mechanism controlling the turning of the key. The keyhole is a narrow twisted slot in the face of a cylinder,G(Fig. 219), which revolves inside a larger fixed cylinder,F. As the key is pushed in, the notches in its upper edge raise up the pinsA1, B1, C1, D1, E1, until their tops exactly reach the surface ofG, which can now be revolved by the key in Fig. 220, and work the bolt through the medium of the armH. (The bolt itself is not shown.) If a wrong key is inserted, either some of the lower pins will project upwards into the fixed cylinderF(see Fig. 221), or some of the pins inFwill sink intoG. It is then impossible to turn the key.

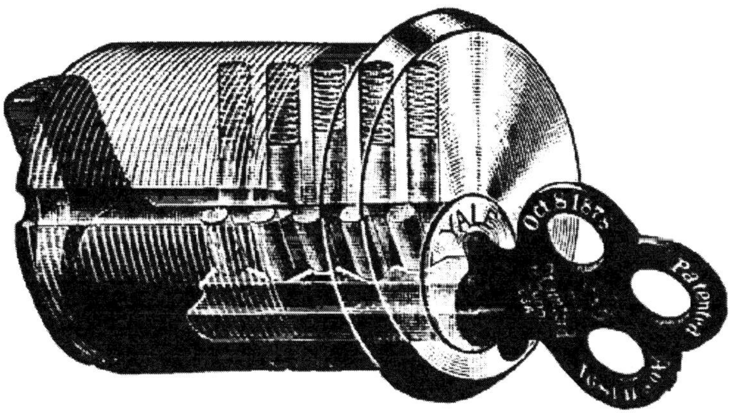

Fig. 220.—Yale key turning.

There are other well-known locks, such as those invented by Bramah and Hobbs. But as these do not lend themselves readily to illustration no detailed account can be given. We might, however, notice the *time* lock, which is set to a certain hour, and can be opened by the right key or a number of keys in combination only when that hour is reached. Another very interesting device is the *automatic combination* lock. This may have twenty or more keys, any one of which can lock it; but the same one must be used to *un*lock it, as the key automatically sets the mechanism in favour of itself. With such a lock it would be possible to have a different key for every day in the month; and if any one key got into wrong hands it would be useless unless it happened to be the one which last locked the lock.

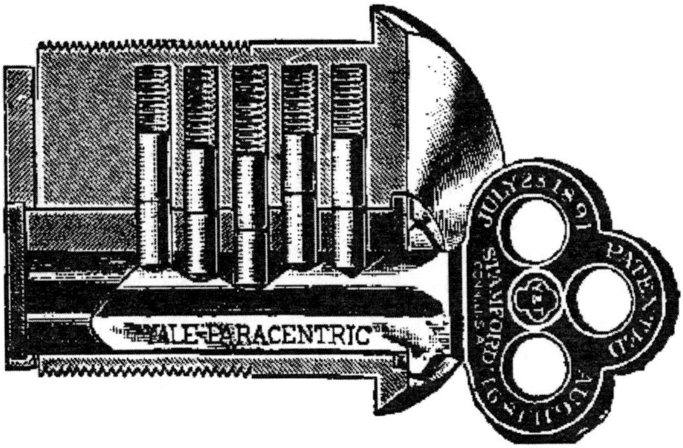

Fig. 221.—The wrong key inserted. The pins do not allow the lock to be turned.

THE CYCLE.

There are a few features of this useful and in some ways wonderful contrivance which should be noticed. First,

THE GEARING OF A CYCLE.

To a good many people the expression "geared to 70 inches," or 65, or 80, as the case may be, conveys nothing except the fact that the higher the gear the faster one ought to be able to travel. Let us therefore examine the meaning of such a phrase before going farther.

The safety cycle is always "geared up"—that is, one turn of the pedals will turn the rear wheel more than once. To get the exact ratio of turning speed we count the teeth on the big chain-wheel, and the teeth on the small chain-wheel attached to the hub of the rear wheel, and divide the former by the latter. To take an example:—The teeth are 75 and 30 in number respectively; the ratio of speed therefore $= 75/30 = 5/2 = 2\frac{1}{2}$. One turn of the pedal turns the rear wheel $2\frac{1}{2}$ times. The gear of the cycle is calculated by multiplying this result by the diameter of the rear wheel in inches. Thus a 28-inch wheel would in this case give a gear of $2\frac{1}{2} \times 28 = 70$ inches.

One turn of the pedals on a machine of this gear would propel the rider as far as if he were on a high "ordinary" with the pedals attached directly to a wheel 70 inches in diameter. The gearing is raised or lowered by altering the number ratio of the teeth on the two chain-wheels. If for the 30-tooth wheel we substituted one of 25 teeth the gearing would be—

$75/25 \times 28$ inches = 84 inches.

A handy formula to remember is, gearing = T/ t × D, where T = teeth on large chain-wheel; t = teeth on small chain-wheel; and D = diameter of driving-wheel in inches.

Two of the most important improvements recently added to the cycle are—(1) The free wheel; (2) the change-speed gear.

THE FREE WHEEL

is a device for enabling the driving-wheel to overrun the pedals when the rider ceases pedalling; it renders the driving-wheel "free" of the driving gear. It is a ratchet specially suited for this kind of work. From among the many patterns now marketed we select the Micrometer free-wheel hub (Fig. 222), which is extremely simple. The *ratchet-wheel*Ris attached to the hub of the driving-wheel. The small chain-wheel (or "chain-ring," as it is often called) turns outside this, on a number of balls running in a groove chased in the neck of the ratchet. Between these two parts are the *pawls,* of half-moon shape. The driving-wheel is assumed to be on the further side of the ratchet. To propel the cycle the chain-ring is turned in a clockwise direction. Three out of the six pawls at once engage with notches in the ratchet, and are held tightly in place by the pressure of the chain-ring on their rear ends. The other three are in a midway position.

Fig. 222.

When the rider ceases to pedal, the chain-ring becomes stationary, but the ratchet continues to revolve. The pawls offer no resistance to the ratchet teeth, which push them up into the semicircular recesses in the chain-ring. Each one rises as it passes over a tooth. It is obvious that driving power cannot be transmitted again to the road wheel until the chain-wheel is turned fast enough to overtake the ratchet.

THE CHANGE-SPEED GEAR.

A gain in speed means a loss in power, and *vice versâ*. By gearing-up a cycle we are able to make the driving-wheel revolve faster than the pedals, but at the expense of control over the driving-wheel. A high-geared cycle is fast on the level, but a bad hill-climber. The low-geared machine shows to disadvantage on the flat, but is a good hill-climber. Similarly, the express engine must have large driving-wheels, the goods engine small driving-wheels, to perform their special functions properly.

In order to travel fast over level country, and yet be able to mount hills without undue exertion, we must be able to do what the motorist does—change gear. Two-speed and three-speed gears are now very commonly fitted to cycles. They all work on the same principle, that of the epicyclic train of cog-wheels, the mechanisms being so devised that the hub turns more slowly than, at the same speed as, or faster than the small chain-wheel,[42] according to the wish of the rider.

We do not propose to do more here than explain the principle of the epicyclic train, which means "a wheel on (or running round) a wheel." Lay a footrule on the table and roll a cylinder along it by the aid of a second rule, parallel to the first, but resting on the cylinder. It will be found that, while the cylinder advances six inches, the upper rule advances twice that distance. In the absence of friction the work done by the agent moving the upper rule is equal to that done in overcoming the force which opposes the forward motion of the cylinder; and as the distance through which the cylinder advances is only half that through which the upper rule advances, it follows that the *force* which must act on the upper rule is only half as great as that overcome in moving the cylinder. The carter makes use of this principle when he puts his hand to the top of a wheel to help his cart over an obstacle.

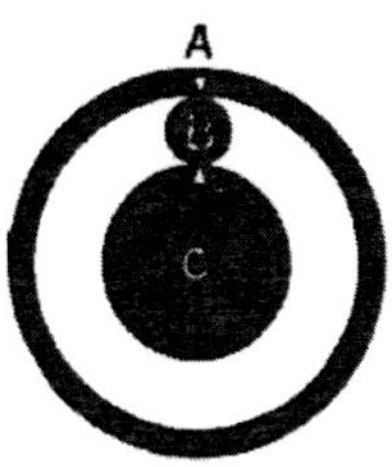

Fig. 223.

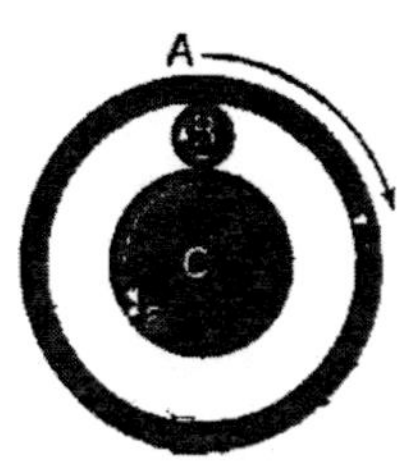

Fig. 224.

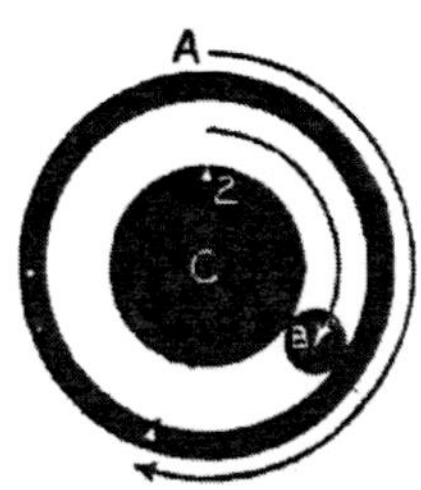

Fig. 225.

Now see how this principle is applied to the change-speed gear. The lower rule is replaced by a cog-wheel,C(Fig. 223); the cylinder by a cog,B, running round it; and the upper rule by a ring,A, with internal teeth. We may suppose thatAis the chain-ring,Ba cog mounted on a pin projecting from the hub, andCa cog attached to the fixed axle. It is evident thatBwill not move so fast roundCasAdoes. The amount by whichAwill get ahead ofBcan be calculated easily. We begin with the wheels in the position shown in Fig. 223. A point,I, onAis exactly over the topmost point ofC. For the sake of convenience we will first assume that instead ofBrunning roundC,Bis revolved on its axis for one complete revolution in a clockwise direction, and thatAandCmove as in Fig. 224. IfBhas 10 teeth,C30, andA40,Awill have been moved10⁄40 = ¼ of a revolution in a clockwise direction, andC10⁄30 = ⅓ of a revolution in an anti-clockwise direction.

Now, coming back to what actually does happen, we shall be able to understand how farArotates roundCrelatively to the motion ofB, whenCis fixed andBrolls (Fig. 225).Badvances ⅓ of distance roundC;Aadvances ⅓ ¼ =7⁄12 of distance roundB. The fractions, if reduced to a common denominator, are as 4:7, and this is equivalent to 40 (number of teeth onA): 40 30 (teeth onAteeth onC.)

To leave the reader with a very clear idea we will summarize the matter thus:—If T = number of teeth onA, *t* = number of teeth onC, then movement ofA: movement ofB:: T *t*: T.

Here is a two-speed hub. Let us count the teeth. The chain-ring (=A) has 64 internal teeth, and the central cog (=C) on the axle has 16 teeth. There are four cogs (=B) equally spaced, running on pins projecting from the hub-shell betweenAandC. How much faster thanBdoesArun roundC? Apply the formula:—Motion ofA: motion ofB:: 64 16: 64. That is, whileArevolves once,Band the hub and the driving-wheel will revolve only64⁄80 = ⅘ of a turn. To use scientific language,Brevolves 20 per cent. slower thanA.

This is the gearing we use for hill-climbing. On the level we want the driving-wheel to turn as fast as, or faster than, the chain-ring. To make it turn at the same rate, bothAandCmust revolve together. In one well-known gear this is effected by slidingCalong the spindle of the wheel till it disengages itself from the spindle, and one end locks with the plate which carriesA. SinceBis now being pulled round at the bottom as well as the top, it cannot rotate on its own axis any longer, and the whole train revolves *solidly*—that is, whileAturns through a circleBdoes the same.

To get an *increase* of gearing, matters must be so arranged that the drive is transmitted from the chain-wheel toB, and fromAto the hub. WhileBdescribes a circle,Aand the driving-wheel turn through a circle and a part of a circle—

that is, the driving-wheel revolves faster than the hub. Given the same number of teeth as before, the proportional rates will beA= 80,B= 64, so that the gear *rises* 25 per cent.

By means of proper mechanism the power is transmitted in a three-speed gear either (1) from chain-wheel toA,AtoB,Bto wheel = *low* gear; or (2) from chain-wheel toAandCsimultaneously = solid, normal, or *middle* gear; or (3) from chain-wheel toB,BtoA,Ato wheel = *high* gear. In two-speed gears either 1 or 3 is omitted.

AGRICULTURAL MACHINES.

THE THRESHING-MACHINE.

Bread would not be so cheap as it is were the flail still the only means of separating the grain from the straw. What the cream separator has done for the dairy industry the threshing-machine has done for agriculture. A page or two ought therefore to be spared for this useful invention.

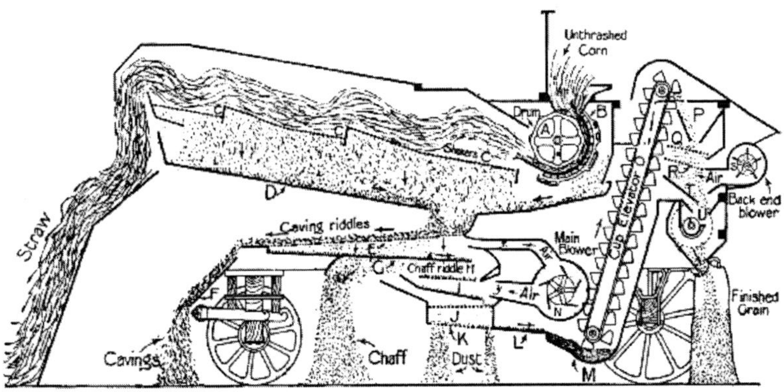

Fig. 226.—Section of a threshing machine.

In Fig. 226 a very complete fore-and-aft section of the machine is given. After the bands of the sheaves have been cut, the latter are fed into the mouth of the *drum*Aby the feeder, who stands in the feeding-box on the top of the machine. The drum revolves at a very high velocity, and is fitted with fluted beaters which act against a steel concave, or breastwork,B, the grain being threshed out of the straw in passing between the two. The breastwork is provided with open wires, through which most of the threshed grain, cavings (short straws), and chaff passes on to a sloping board. The straw is flung forward on to the shakersC, which gradually move the straw towards the open end and throw it off. Any grain, etc., that has escaped the drum falls through the shakers on toD, and works backwards to the *caving riddles*, or moving sieves,E. The *main blower*, by means of a revolving fan,N, sends air along the channelXupwards through these riddles, blowing the short straws away to the left. The grain, husks, and dust fall throughEon toG, over the end of which they fall on to the *chaff riddle*,H. A second column of air from the blower drives the chaff away. The heavy grain, seeds, dust, etc., fall on toI,J, andKin turn, and are shaken until only the grain remains to pass alongLto the elevator bottom,M. An endless band with cups attached to it scoops up the grain, carries it aloft, and shoots it into hopperP. It then goes through the shakersQ, R, is dusted by the *back end blower*,S, and slides downTinto the open end of the rotary screen-drumU, which is mounted on the slope, so that as it turns the grain travels gradually along it. The first half of the screen has wires set closely together. All the small grain that falls through this, called "thirds," passes into a hopper, and is collected in a sack attached to the hopper mouth. The "seconds" fall through the second half of the drum, more widely spaced, into their sack; and the "firsts" fall out of the end and through a third spout.

MOWING-MACHINES.

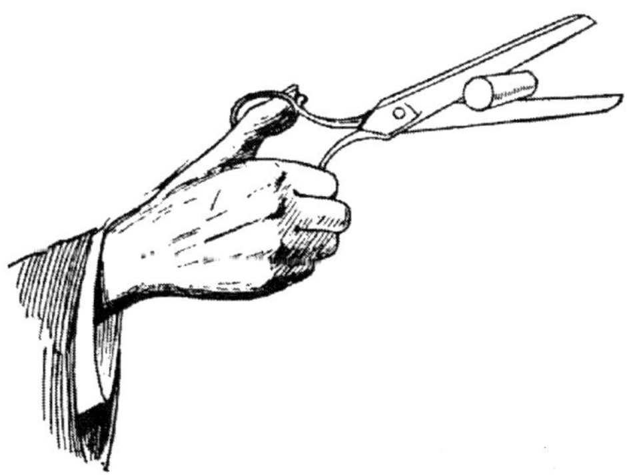

Fig. 227.

The ordinary *lawn—mower* employs a revolving reel, built up of spirally-arranged knives, the edges of which pass very close to a sharp plate projecting from the frame of the mower. Each blade, as it turns, works along the plate, giving a shearing cut to any grass that may be caught between the two cutting edges. The action is that of a pair of scissors (Fig. 227), one blade representing the fixed, the other the moving knife. If you place a cylinder of wood in the scissors it will be driven forward by the closing of the blades, and be marked by them as it passes along the edges. The same thing happens with grass, which is so soft that it is cut right through.

HAY-CUTTER.

The *hay-cutter* is another adaptation of the same principle. A cutter-bar is pulled rapidly backwards and forwards in a frame which runs a few inches above the ground by a crank driven by the wheels through gearing. To the front edge of the bar are attached by one side a number of triangular knives. The frame carries an equal number of spikes pointing forward horizontally. Through slots in these the cutter-bar works, and its knives give a drawing cut to grass caught between them and the sides of the spikes.

SOME NATURAL PHENOMENA.

WHY SUN-HEAT VARIES IN INTENSITY.

The more squarely parallel heat-rays strike a surface the greater will be the number that can affect that surface. This is evident from Figs. 228, 229, whereA Bis an equal distance in both cases. The nearer the sun is to the horizon, the more obliquely do its rays strike the earth. Hence midday is necessarily warmer than the evening, and the tropics, where the sun stands overhead, are hotter than the temperate zones, where, even in summer at midday, the rays fall more or less on the slant.

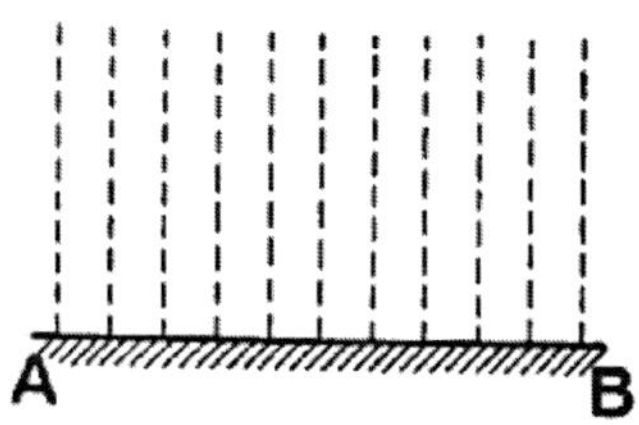

Fig. 228.

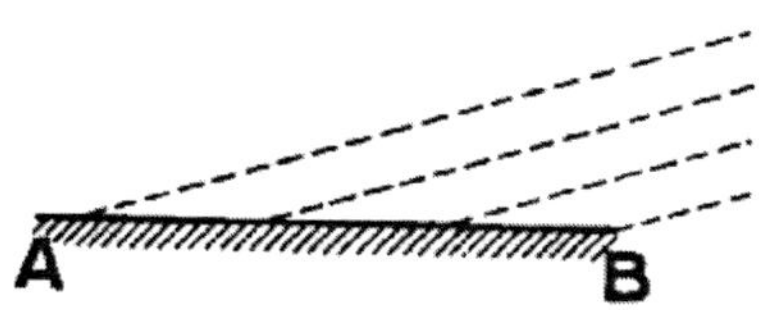

Fig. 229.

The atmospheric envelope which encompasses the earth tends to increase the effect of obliquity, since a slanting ray has to travel further through it and is robbed of more heat than a vertical ray.

THE TIDES.

All bodies have an attraction for one another. The earth attracts the moon, and the moon attracts the earth. Now, though the effect of this attraction is not visible as regards the solid part of the globe, it is strongly manifested by the water which covers a large portion of the earth's surface. The moon attracts the water most powerfully at two points, that nearest to it and that furthest away from it; as shown on an exaggerated scale in Fig. 230. Since the earth and the water revolve as one mass daily on their axis, every point on the circumference would be daily nearest to and furthest from the moon at regular intervals, and wherever there is ocean there would be two tides in that period, were the moon stationary as regards the earth. (It should be clearly understood that the tides are not great currents, but mere thickenings of the watery envelope. The inrush of the tide is due to the temporary rise of level.)

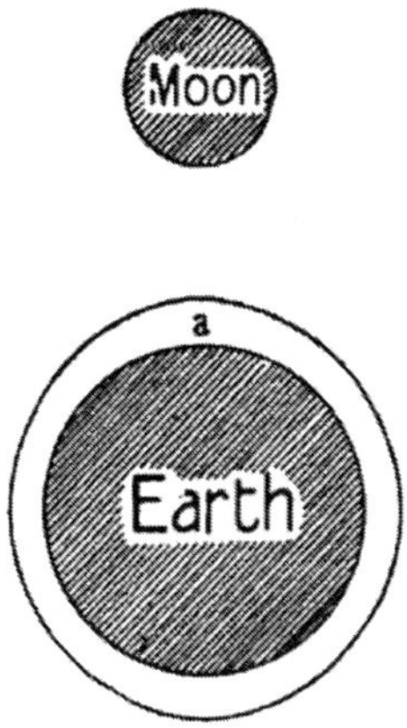

Fig. 230.

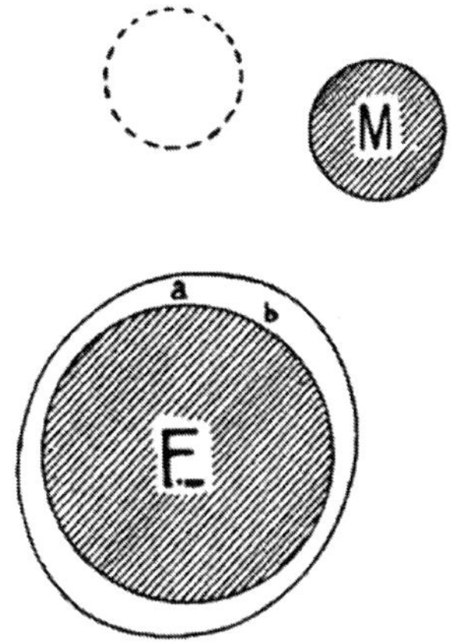

Fig. 231.

WHY HIGH TIDE VARIES DAILY.

The moon travels round the earth once in twenty-eight days. In Fig. 231 the point *a* is nearest the moon at, say, twelve noon. At the end of twenty-four hours it will have arrived at the same position by the compass, but yet not be nearest to the moon, which has in that period moved on1/28th of a revolution round the earth [43] Consequently high tide will not occur till *a* has reached position *b* and overtaken the moon, as it were, which takes about an hour on the average. This explains why high tide occurs at intervals of more than twelve hours.

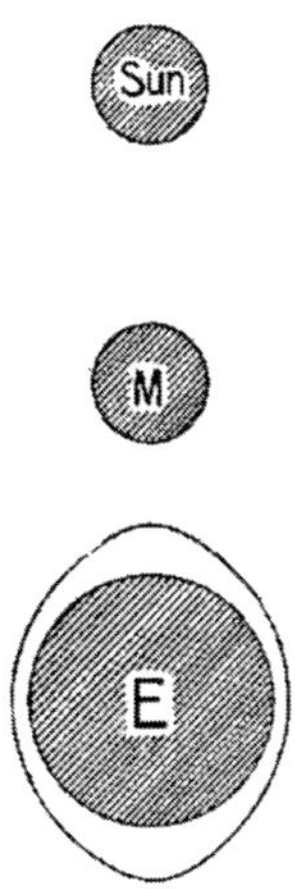

Fig. 232.—Relative positions of sun, moon, and earth at "spring" tides.

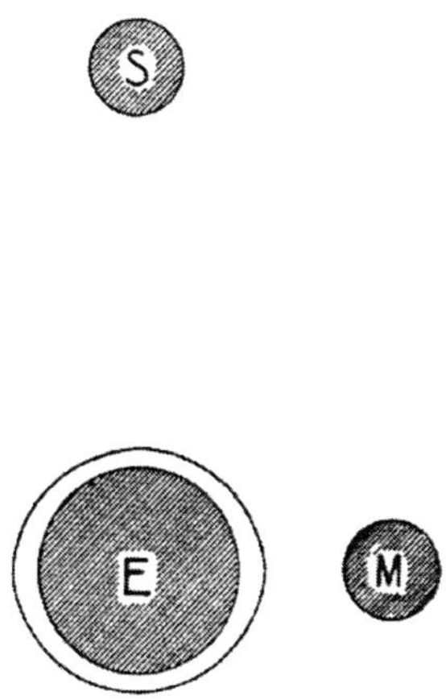

Fig. 233.—Relative positions of sun, moon, and earth at "neap" tides.

NEAP TIDES AND SPRING TIDES.

The sun, as well as the moon, attracts the ocean, but with less power, owing to its being so much further away. At certain periods of the month, sun, earth, and moon are all in line. Sun and moon then pull together, and we get the highest, or *spring* tides (Fig. 232). When sun and moon pull at right angles to one another—namely, at the first and third quarters—the excrescence caused by the moon is flattened (Fig. 233), and we get the lowest, or *neap* tides.

[39] In both Figs. 207 and 208 the degree of expansion is very greatly exaggerated.

[40] As the sun passes the meridian (twelve o'clock, noon) the chronometer's reading is taken, and the longitude, or distance east or west of Greenwich, is reckoned by the difference in time between local noon and that of the chronometer.

[41] For much of the information given here about clocks and watches the author is indebted to "The History of Watches," by Mr. J.F. Kendal.

[42] We shall here notice only those gears which are included in the hub of the driving-wheel.

[43] The original position of the moon is indicated by the dotted circle.

INDEX.